Applications of Advanced Optimization Techniques in Industrial Engineering

Information Technology, Management and Operations Research Practices

Series Editors: Vijender Kumar Solanki, Sandhya Makkar, and Shivani Agarwal

This new book series will encompass theoretical and applied books and will be aimed at researchers, doctoral students, and industry practitioners to help in solving real-world problems. The books will help in the various paradigm of management and operations. The books will discuss the concepts and emerging trends on society and businesses. The focus is to collate the recent advances in the field and take the readers on a journey that begins with understanding the buzz words like employee engagement, employer branding, mathematics, operations, technology and how they can be applied in various aspects. It walks readers through engaging with policy formulation, business management, and sustainable development through technological advances. It will provide a comprehensive discussion on the challenges, limitations, and solutions of everyday problems like how to use operations, management and technology to understand the value-based education system, health and global warming, and real-time business challenges. The book series will bring together some of the top experts in the field throughout the world who will contribute their knowledge regarding different formulations and models. The aim is to provide the concepts of related technologies and novel findings to an audience that incorporates specialists, researchers, graduate students, designers, experts, and engineers who are occupied with research in technology, operations, and management related issues.

Performance Management
Happiness and Keeping Pace with Technology
Edited by Madhu Arora, Poonam Khurana, and Sonam Choiden

Soft Computing Applications and Techniques in Healthcare
Edited by Ashish Mishra, G. Suseendran, and Trung-Nghia Phung

Employer Branding for Competitive Advantage
Models and Implementation Strategies
Edited by Geeta Rana, Shivani Agarwal, and Ravindra Sharma

Analytics in Finance and Risk Management
Edited by Sweta Agarwal, Nidhi Malhotra, and T. P. Ghosh

Quality of Life
An Interdisciplinary Perspective
Edited by Shruti Tripathi, Rashmi Rai, and Ingrid Van Rompay-Bartels

Applications of Advanced Optimization Techniques in Industrial Engineering
Edited by Abhinav Goel, Anand Chauhan, and A.K. Malik

For more information about this series, please visit: https://www.routledge.com/Information-Technology-Management-and-Operations-Research-Practices/book-series/CRCITMORP

Applications of Advanced Optimization Techniques in Industrial Engineering

Edited by
Abhinav Goel
Anand Chauhan
A.K. Malik

CRC Press
Taylor & Francis Group
Boca Raton London New York

CRC Press is an imprint of the
Taylor & Francis Group, an **informa** business

First edition published 2022
by CRC Press
6000 Broken Sound Parkway NW, Suite 300, Boca Raton, FL 33487-2742

and by CRC Press
2 Park Square, Milton Park, Abingdon, Oxon, OX14 4RN

Library of Congress Cataloging-in-Publication Data
Names: Goel, Abhinav, editor. | Chauhan, Anand, editor. | Malik, A. K., editor.
Title: Applications of advanced optimization techniques in industrial
engineering / edited by Abhinav Goel, Anand Chauhan, and A.K. Malik.
Description: First edition. | Boca Raton : CRC Press, [2022] | Series:
Information technology, management and operations research practices |
Includes bibliographical references and index.
Identifiers: LCCN 2021044503 (print) | LCCN 2021044504 (ebook) | ISBN
9780367545451 (hbk) | ISBN 9780367545468 (pbk) | ISBN 9781003089636 (ebk)
Subjects: LCSH: Industrial efficiency. | Engineering economy. | Mathematical optimization.
Classification: LCC T58.8 .A67 2022 (print) | LCC T58.8 (ebook) | DDC 658.5--dc23
LC record available at https://lccn.loc.gov/2021044503
LC ebook record available at https://lccn.loc.gov/2021044504

ISBN: 978-0-367-54545-1 (hbk)
ISBN: 978-0-367-54546-8 (pbk)
ISBN: 978-1-003-08963-6 (ebk)

DOI: 10.1201/9781003089636

Typeset in Times
by SPi Technologies India Pvt Ltd (Straive)

Contents

Editors

Dr. Abhinav Goel is an assistant professor of mathematics at Graphic Era (Deemed to be University) Dehradun, Uttarakhand, India. He earned his MSc degree in mathematics from Chaudhary Charan Singh University, Meerut, Uttar Pradesh, India and his PhD from Hemwati Nandan Bahuguna (A Central University) University, Srinagar Uttarakhand, India. He has 11 years' experience teaching in academic and research institutions. He has taught at many reputed institutes in the National Capital Region, including ITS Ghaziabad and RKGIT Ghaziabad. His teaching interests are in the areas of Engineering Mathematics, Applied Mathematics, Operations Research, and Numerical Techniques, among others. His area of specialization is Operations Research, Inventory Control, Genetic Algorithm and Soft Computing techniques. He has also attended and presented his research papers at various national and international seminars/conferences/symposiums/workshops/Faculty Development Programs. He has completed a certification course from HarvardX on Data Science. He is a reviewer and editorial board member of various reputed national and international journals. He is an author of four books, including *Engineering Mathematics for GATE Entrance* (published by Arihant Publication, India). Currently he is guiding three PhD students.

Dr. Anand Chauhan is an associate professor and Head of the Department of Mathematics, Graphic Era (Deemed to be University), Dehradun, India. He completed a Master of Science in Mathematics from H.N.B Garhwal Central University, Srinagar, Uttarakhand and earned a PhD in Mathematics from Chaudhary Charan Singh University, Meerut, Uttar Pradesh, India. He has more than 12 years of teaching and research experience. Over the course of his academic career, he has published more than 30 research papers in national and international peer-reviewed journals and he has attended more than 24 national and international conferences/seminars/workshops. Dr. Chauhan is actively involved in research fields including operational research, inventory control, optimization, numerical methods, and he has teaching interests in Engineering Mathematics, Discrete Mathematics, Graph Theory, Operational Research, and Computer-Based Numerical and Statistical Technique. Three PhDs have been awarded under his supervision and he has also been an organizing secretary of three international conferences. He is an author/co-author of more than seven books, a co-editor of the *International Journal of Operations Research and Optimization* and also works as a reviewer for many national and international journals. He is a life member of the Indian Mathematical Society (IMS) and the Indian Science Congress Association (ISCA).

Dr. A. K. Malik is an associate professor of mathematics at the B K Birla Institute of Engineering & Technology, Pilani, Rajasthan, India. He earned his MSc degree in mathematics from Gurukula Kangri Vishwavidyalaya, Haridwar, Uttrakhand and his PhD from Chaudhary Charan Singh University, Meerut, Uttar Pradesh, India. He has 14 years' experience teaching in academic and research institutions. He was presented with the Srinivasa Ramanujan Award for Mathematics Popularization, on December 22, 2018 on National Mathematics Day from VIPNET-Vigyan Prasar (under DST Govt. of India) Network DASA & SWA of Tripura. His areas of specialization are Operations Research, Inventory Control, and Soft Computing techniques. He has published more than fifty research papers in reputed national and international journals, and he has presented his research papers at various national and international seminars/conferences/symposiums/workshops. He is the author/co-author of 19 books, including *Operations Research* (Oxford University Press), *A Textbook of Engineering Mathematics-II* (Manakin Press, New Delhi), *Engineering Mathematics-I & II, Mathematics-III* (for EC&EE*), Advanced Engineering Mathematics* (All New Age International Publication, New Delhi); *Optimization Techniques, Topology, Measure Theory & Integration* (I K International Publication, New Delhi); and *Mathematics-III (for CS& IT)* (Asian Publication, New Delhi) etc. He has been invited to deliver talks on various research topics in several institutes and universities, such as IITs.

Contributors

Megha Agarwal
SRM IST
Ghaziabad, India

Mohd Aftab Ali
Department of Mathematics
Shri Venkateshwara University
Gajraula

Irfan Ali
Department of Statistics & Operations Research
Aligarh Muslim University
Aligarh, India

Ashish
Department of Mathematics
Government College Satnali
Mahendergarh, Haryana, India

Ramakant Bhardwaj
Department of Mathematics
Amity University Kolkata
India (W.B.) India

Mohamed Boualem
Research Unit LaMOS (Modeling and Optimization of Systems)
Faculty of Technology
University of Bejaia
Bejaia, Algeria

Amina Angelika Bouchentouf
Laboratory of Mathematics
Djillali Liabes University of Sidi Bel Abbes
Algeria

Anand Chauhan
Graphic Era Deemed to be University
Dehradun, India

Shail Kr Dinkar
Department of Computer Science and Applications
G. B. Pant Institute of Engineering and Technology
Pauri Garhwal, Uttarakhand, India

Abhinav Goel
Department of Mathematics
Graphic Era University
Dehradun, India

Srikant Gupta
Jaipuria Institute of Management
Jaipur, Rajasthan, India

Vinti Gupta
SRM IST
Ghaziabad, India

Javid Iqbal
Department of Mathematical Sciences
BGSB University
Rajouri, Jammu and Kashmir, India

Ajay Jha
Department of Transportation Management
University of Petroleum and Energy Studies
Dehradun, India

Bhuwan Chandra Joshi
Graphic Era Deemed to be University
Dehradun, India

Pankaj Kumar Jha
Department of Statistic
Amity University
Kolkata, India

R. Karnwal
Jaypee Institute of Information Technology
Noida, India

Khamosh
Department of Mathematics
Baba Mastnath University
Rohtak, Haryana,

Amit Kumar
Department of Mathematics
Chandigarh University
Mohali (Punjab), India

Dr. Rakesh Kumar
Department of Applied Sciences
Shaheed Bhagat Singh State University
Ferozepur, Punjab, India

Vimal Kumar
Department of Information Management
Chaoyang University of Technology
Taichung, Taiwan

A. K. Malik
Department of Applied Science and Humanities
B K Birla Institute of Engineering & Technology
Pilani, Rajasthan

R. Pal
Jaypee Institute of Information Technology
Noida, India

Neetu Paliwal
Department of Physical Science, RNTU
Bhopal, India

R. K. Pandey
D. B. S. (PG) College,
Dehradun, India

Abhijit Pandit
Amity University
Kolkata, India

Praveen Kumar Poonia
Department of General Requirements,
Ibri College of Applied Sciences
Oman

Ather Aziz Raina
Department of Mathematics,
Govt. Degree College Darhal,
Rajouri, Jammu and Kashmir, India

Neelanjana Rajput
H.N.B.G.U.
Srinagar Garhwal
Uttarakhand, India

Savita
Department of Mathematics
Chandigarh University
Mohali (Punjab), India

Archana Sharma
Department of Applied Sciences
KIET Group of Institutions
Delhi-NCR, Ghaziabad, U.P.

Chandra Mani Sharma
School of Computer Science
University of Petroleum & Energy Studies
Uttarakhand, India

Nagendra Kumar Sharma
Department of Business Administration
Chaoyang University of Technology
Taichung, Taiwan

Anu Sirohi
Department of Mathematics
Bharat Institute of Technology
UP, India

Ritu Saxena
Department of Physical Science
RNTU
Bhopal, India

Pratima Verma
Department of Information Management
Chaoyang University of Technology
Taichung, Taiwan

S. Yadav
Jaypee Institute of Information Technology,
Noida, India

1 Dynamical Analysis in Modulated Logistic Maps

Ashish, A. K. Malik, and Khamosh

CONTENTS

1.1 INTRODUCTION

In the twenty-first century, nonlinear science and its dynamics has played a crucial role in the various branches of science, such as the population growth model, weather forecasting, stock market fluctuations, the motion of the stars and galaxies, communication security, transportation problems, etc. The dynamics of such types of time evolutionary systems are determined using difference and differential equations. Therefore, the logistic equation $\mu x(1-x)$ is considered as one of the famous difference equations, which was first introduced in 1838 by P. F. Verhulst. In 1976, May [1] and Oster et al. [2] introduced several difference maps and established their applications in various branches of science. For example, the Ricker map $x\, exp\mu^{(1-x)}$ is assumed as a model of population growth of single species for epidemic diseases and the equation $\mu x(1-x)^{-b}$ is used to examine the population of insects. In 1978, the dynamical behavior of the logistic map was popularized with a mathematical model of chaos given by Feigenbaum [3]. Further, for the brief study on the dynamics of one-dimensional maps one may refer to Devaney [4, 5], Holmgren [6], Block and Coppel [7], Wiggins [8], Alligood et al. [9], Ausloos and Dirickx [10], Elagdi [11], Martelli [12], Strogatz [13], etc.

Over the years, the chaotic behavior, period-doubling bifurcation, period-3 implies chaos and Lyapunov exponent were studied using various forms of the standard logistic map, such as the delayed logistic map [14], generalized form $\mu x^p(1-x)^q$ with stability of its solutions [15], the spatial logistic map for spatiotemporal chaos [16],

DOI: 10.1201/9781003089636-1

kicked logistic map with freedom of three parameters [17], the phase-modulated logistic map [18], the logistic map with modulated parameter [19], the fractional logistic map with chaos [20] and generalized logistic maps with changes in the parameter sign [21, 22]. Further, instead of generalized modulated logistic maps, various new techniques were introduced to examine the dynamical behavior of the logistic maps. For example, Ashish et al. [23], using superior feedback systems, established the chaotic behavior of standard logistic map for the growth- rate parameter $\mu \in [0, 4.22]$. In addition, they established analytical properties such as fixed and periodic state behavior of the logistic map using the superior feedback technique [24, 25]. In 2009, Rani and Agarwal [26] studied the periodic and stable behavior of the logistic map using the Mann iterative process for larger range of the parameter μ.

Also, the one-dimensional difference maps have various applications in every branch of science and nature, such as biology, physics, chemistry, engineering, etc. In 2009, Kanso and Smaoui [27] used a logistic map to generate the random numbers (see also [28]). The logistic map was used in cryptography to encrypt and decrypt images, text messages and videos by many mathematicians, such as Pareek et al. [29], Baptista [30], Hafiz et al. [31] and Wang et al. [32]. Singh and Sinha [33] used the logistic map as a tool to introduce the secure communication system and produced the chaotic signals using a logistic system. In 2008, Salarich and Alasty [34] used it as a special case to annihilate chaos in the entropy control technique. In 2009, Medina et al. [35] used the logistic map as a method to define analog generators of chaotic noise. For more applications on chaos theory, one may also refer to Diamond [36], Robinson [37], Sharkovsky [38], etc.

This chapter is divided into seven sections. This initial section (Section 1.1) presents a short literature review on the dynamics of one-dimensional maps and Section 1.2 includes some preliminary results which are used in further sections. Further, in Sections 1.3–1.6, the dynamical characteristics of cubic and quartic type modulated logistic map for various values of parameters p and q are established. Finally, the chapter is concluded in Section 1.7.

1.2 PRELIMINARIES

Here, in this section, we deal with the basic entities which are used in further sections to understand the dynamical characteristics of one-dimensional maps.

Definition 1.1. Let $x \in X$be a point in a one-dimensional map f, where X is a non-empty set then x is said to be fixed if $f(x) = x$ and is also said to be periodic of period p if $f^p(x) = x$, where p is a positive integer [4].

Definition 1.2. Let $x \in X$ be a fixed point in a one-dimensional map f, where X is a non-empty set then, x is said to be attracting if $|f'(x)| < 1$, and said to be repelling if $|f'(x)| > 1$. Further, the point x is said to be neutral if $|f'(x)| = 1$ [4].

Definition 1.3. A point x in a one-dimensional system f is known as critical if $f'(x) = 0$.

Definition 1.4. Let $\{x_p\}$ be an iterative sequence of period-p for the one-dimensional map f, then the first order derivative of pth iterate of f is defined as:

$$(f_p)'(x_1) = f'(x_p).f'(x_{p-1}).....f'(x_2).f'(x_1)$$

and is known as the chain rule of product for periodic fixed points [4].

1.3 MODULATED LOGISTIC MAPS

The discrete one-dimensional maps are assumed essential in modelling and the dynamics of nonlinear phenomena. Therefore, in this section, we start by introducing the following modulated logistic map:

$$M_{p,q}(\mu, x) = \mu x^p (1-x)^q, (say) \tag{1.1}$$

where $p, q > 0$, $x \in [0, 1]$ and $\mu \in (0, \mu_{max}]$. It is noticed that the modulated logistic map reduces into standard logistic map $\mu x(1 - x)$ at $(p, q) = (1, 1)$, where $\mu \in (0, 4]$. Also, for the starter $x_0 \in [0, 1]$ the above relation (1.1) reduces into the following difference equation:

$$x_{n+1} = \mu x_n^p (1-x_n)^q \tag{1.2}$$

where $x_n \in [0, 1]$, $n = 0, 1, 2, \ldots$ and the sequence $\{x_0, x_1, x_2, \ldots x_n\}$ established in (1.2) is known as an iterative orbit of the modulated logistic map, where the system admits all its dynamical characteristics. Further, for the modulated logistic system (1.1) it is assumed that the nth iterate $M^n(\mu, x) \in [0, 1]$ for each value of the growth-rate parameter $\mu \in (0, \mu_{max}]$. Furthermore, as the dynamical behavior of the original map depends on the parameters p, q and μ, therefore, for the arbitrary values of parameters p and q the system (1.1) may generate lot of novel one-dimensional maps which can be examined mathematically as well as analytically. Thus, for the sake of simplicity we take three cases $(p, q) = (2, 1)$, $(1, 2)$ and $(2, 2)$ and obtain their dynamical properties. Also, just like the standard logistic map the following interesting results are obtained from the modulated logistic map (1.1) using Definitions 1.1, 1.2 and 1.3:

(a) The critical point, an interesting property of one-dimensional systems, is determined for the modulated logistic map $M_{p,q}(\mu, x)$ by using Definition 1.3 in the following form:

$$M'_{p,q}(\mu,x) = \mu x^{p-1}(1-x)^q - \mu q x^p (1-x)^{q-1} = 0. \tag{1.3}$$

Then, solving Equation (1.3), we get $x = \frac{p}{p+q}$ as the critical point for the modulated logistic map (1.1), where $p, q > 0$, which determines the maximum value of the map.

(b) Since the modulated logistic map $M_{p,q}(\mu, x)$ gives its maximum value 1 at $= \frac{p}{p+q}$. Therefore, the maximum of the growth-rate parameter μ, that is, μ_{max}, is obtained by solving the following relation:

$$M_{p,q}(\mu,x) = \mu x^p (1-x)^q = 1. \tag{1.4}$$

Taking $x = \frac{p}{p+q}$ in Equation (1.4), we find

$$\mu_{max}(p,q) = \frac{(p+q)^{p+q}}{p^p . q^q}. \tag{1.5}$$

(c) For $p, q > 0$, the physical fixed-point solutions of a one-dimensional system $M_{p,q}(\mu, x)$ using Definition 1.1, are obtained in the following way:

$$M_{p,q}(\mu,x) = \mu x^p (1-x)^q = x \tag{1.6}$$

where $x \in [0, 1]$ and $\mu \in (0, \mu_{max}]$.

1.4 DYNAMICAL ANALYSIS IN $\mu x^p(1 - x)^q$ FOR $p = 2$ AND $q = 1$

In the following section of this chapter the dynamical properties, such as peak value, critical point, fixed point, periodic point and chaotic behavior of the modulated logistic map in for $p = 2$ and $q = 1$, are studied. Therefore, we assume the following form of the modulated logistic map $M_{p,q}(\mu, x)$:

$$M_{2,1}(\mu,x) = \mu x^2 (1-x). \tag{1.7}$$

Then, using Definition 1.3, for critical point and the relation (1.3) we find $x = \frac{2}{3}$ as the critical point for the cubic-type modulated logistic map (1.7) and hence using (1.5) the maximum value of parameter μ, that is, μ_{max} for the map (1.7) is equal to 6.75. Therefore, all of the dynamical behavior of the modulated map $M_{2,1}(\mu, x)$ occurs in the interval (0, 6.75]. Further, the functional graph of the cubic-type modulated logistic map is illustrated by the family of parabolic curves which, at $x = \frac{2}{3}$, attains its maximum value one as shown in Figure 1.1. Therefore, first we determine the fixed-point property and then examine the complete dynamical behavior using period-doubling bifurcation plot.

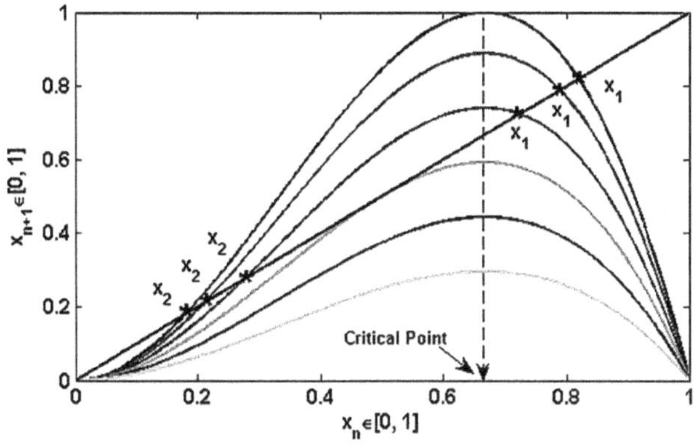

FIGURE 1.1 Functional plot for the cubic type modulated logistic map $M_{2,1}(\mu, x)$ for $0 \leq \mu \leq 6.75$.

Theorem 1.1

Let $M_{2,1}(\mu, x) = \mu x^2(1 - x)$ be the cubic-type modulated logistic map on [0, 1], where $\mu > 0$. Then, show that the points 0 and $\frac{\mu \pm \sqrt{\mu^2 - 4\mu}}{2\mu}$ are the fixed point for the map $M_{2,1}(\mu, x)$, for all $\mu > 0$ and $x \in [0, 1]$.

Proof. Let $M_{2,1}(\mu, x) = \mu x^2(1 - x)$ be the cubic-type modulated logistic map then, using Definition 1.1, for the fixed-point x we have

$$\begin{aligned} \mu x^2(1-x) &= x, \\ x\left(\mu x^2 - \mu x + 1\right) &= 0, \\ \text{either } x = 0 \text{ or } x_1 &= \frac{\mu + \sqrt{\mu^2 - 4\mu}}{2\mu} \text{ and } x_2 = \frac{\mu - \sqrt{\mu^2 - 4\mu}}{2\mu}. \end{aligned} \tag{1.8}$$

Thus, the point $x = 0$, x_1 and x_2 for $\mu > 0$ are the required fixed point. Further, for more simplicity Figure 1.2 shows the complete behavior of the fixed points. It is noticed that for $\mu < 4$ the functional graph of the map $M_{2,1}(\mu, x)$ lies below the diagonal axis, which means origin is the only fixed point, for $\mu = 4$ the functional plot gets more parabolic making a tangent to the diagonal axis and as the value of parameter μ increases through 4, that is, $\mu > 4$ the functional graph intersect the diagonal axis at two fixed points $x_1 = \frac{\mu + \sqrt{\mu^2 - 4\mu}}{2\mu}$ and $x_2 = \frac{\mu - \sqrt{\mu^2 - 4\mu}}{2\mu}$. Thus, it is interesting to know that the cubic-type modulated logistic map bifurcates into two fixed points, x_1 and x_2.

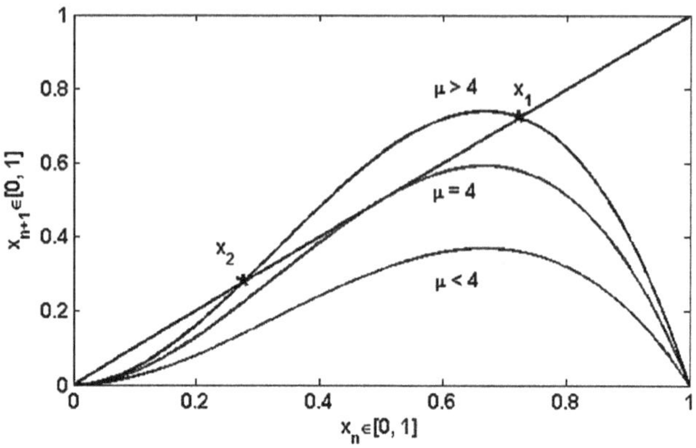

FIGURE 1.2 Functional plot for the cubic type modulated logistic map $M_{2,1}(\mu, x)$ for $\mu < 4$, $\mu = 4$ and $\mu > 4$.

Theorem 1.2

Let $M_{2,1}(\mu, x) = \mu x^2(1 - x)$ be the cubic-type modulated logistic map defined on [0, 1], where $\mu \in [0, 6.75]$. Then, show that $M_{2,1}(\mu, 0) = 0 = M_{2,1}(\mu, 1)$.

Proof. Let $M_{2,1}(\mu, x) = \mu x^2(1 - x)$ be the cubic-type modulated logistic map, then clearly by substituting $x = 0$, we get $M_{2,1}(\mu, x) = 0$ and also taking $x = 1$, we have $M_{2,1}(\mu, 1) = 0$. Figures 1.1 and 1.2, shows the graphical representation for $M_{2,1}(\mu, 0) = 0 = M_{2,1}(\mu, 1)$. This completes the proof.

Theorem 1.3

Let $M_{2,1}(\mu, x) = \mu x^2(1 - x)$ be the cubic-type modulated logistic map on [0, 1] and $\mu \in [0, 6.75]$. Then, the fixed-point $x = 0$ is stable when $0 \le r \le 6.75$ and $x = \frac{\mu + \sqrt{\mu^2 - 4\mu}}{2\mu}$ is stable when $4 \le r \le 5.33$ for the map $M_{2,1}(\mu, x)$.

Proof. To examine the stabilization of the fixed points $x = 0$ and $x = \frac{\mu + \sqrt{\mu^2 - 4\mu}}{2\mu}$, we use Definition 1.1, for the stability of fixed points and prove that $\left|M'_{2,1}(\mu, x)\right| < 1$. Therefore, let us consider

$$\left|M'_{2,1}(\mu, x)\right| = \left|2\mu x - 3\mu x^2\right|. \tag{1.9}$$

Putting $x = 0$ in the relation (1.9), we obtain $\left|M'_{2,1}(\mu, x)\right| < 1$, that means the fixed point 0 is stable for each $\mu \in [0, 6.75]$. Now, substituting the fixed point $x = \frac{\mu + \sqrt{\mu^2 - 4\mu}}{2\mu}$ in the relation (1.9), then, we have

$$\left|M_{2,1}'\left(\mu,\ \frac{\mu+\sqrt{\mu^2-4\mu}}{2\mu}\right)\right| = \left|2\mu\left(\frac{\mu+\sqrt{\mu^2-4\mu}}{2\mu}\right)-3\mu\left(\frac{\mu+\sqrt{\mu^2-4\mu}}{2\mu}\right)^2\right|,$$

$$\left|M_{2,1}'\left(\mu,\ \frac{\mu+\sqrt{\mu^2-4\mu}}{2\mu}\right)\right| = \left|\left(\mu+\sqrt{\mu^2-4\mu}\right)-3\mu\left(\frac{\mu+\sqrt{\mu^2-4\mu}}{2\mu}\right)^2\right|,$$

$$\text{that is, } \left|M_{2,1}'\left(\mu,\ \frac{\mu+\sqrt{\mu^2-4\mu}}{2\mu}\right)\right| < 1,$$

for each $4 \le \mu \le 5.33$. Hence, the fixed point $x = \frac{\mu+\sqrt{\mu^2-4\mu}}{2\mu}$ is also stable. This completes the proof.

Remark 1.4

It is noticed that the fixed point 0 is stable for all $0 \le \mu \le 6.75$ and hence it is a super stable fixed point. Also, the fixed point $x_1 = \frac{\mu+\sqrt{\mu^2-4\mu}}{2\mu}$ is completely stable for $4 \le \mu \le 5.33$ and $x_2 = \frac{\mu-\sqrt{\mu^2-4\mu}}{2\mu}$ is unstable for $4 \le \mu \le 5.33$.

Example 1.5

Let $\mu x^2(1 - x)$ be the cubic-type modulated logistic map on [0, 1] for $\mu \in [0, 6.75]$. Then, prove that for $\mu = 5$ the fixed point $x_1 = 0.723$ is stable and $x_2 = 0.233$ is unstable.

Solution

Using the Definition 1.2, for the stability of fixed point of x_1 we will show that $\left|M_{2,1}'(\mu,x_1)\right| < 1$. Then, we have

$$\left|M_{2,1}'(\mu,x_1)\right| = \left|2\mu x_1 - 3\mu x_1^2\right|, \tag{1.10}$$

where $\mu \in [0, 6.75]$. Then, substituting the fixed-point $x_1 = 0.723$ and $\mu = 5$ in the relation (1.10) then we get

$$\left|M_{2,1}'(5,0.723)\right| = \left|2\times5\times0.723-3\times5\times(0.723)^2\right|,$$

$$= \left|7.23-7.84\right|,$$

$$\text{that is, } \left|M'_{2,1}(5{,}0.723)\right| = \left|-0.61\right| < 1.$$

Thus, the fixed point $x_1 = 0.723$ is stable at $\mu = 5$. Now, to examine the instability for $x_2 = 0.233$ using Definition 1.2 we have to prove that $\left|M'_{2,1}(\mu, x_2)\right| > 1$. Then, again taking $x_2 = 0.233$ and $\mu = 5$ in the relation (1.10), we have

$$\left|M'_{2,1}(5{,}0.233)\right| = \left|2\times5\times0.233 - 3\times5\times(0.233)^2\right|, \quad = \left|2.33 - 0.81\right|,$$

$$\textit{that is, } \left|M'_{2,1}(5{,}0.233)\right| = \left|1.52\right| > 1.$$

Hence, the fixed point $x_2 = 0.233$ is unstable at $\mu = 5$.

1.4.1 Period-Doubling Bifurcation Analysis for the Map $M_{2,1}(\mu, x)$

For $p = 2$ and $q = 1$ the modulated logistic map $\mu x^p(1 - x)^q$ reduces into the cubic-type logistic map $\mu x^2(1 - x)$, where $x \in [0, 1]$ and $\mu \in [0, 6.75]$. Figures 1.3–1.6, shows the complete dynamical behavior for the map $\mu x^2(1 - x)$ for $\mu \in [0, 6.75]$. Therefore, Figure 1.3 shows that there exist two intervals for the growth-rate parameter μ, that is, [0, 4) and [4, 6.75]. For $\mu \in [0, 4)$ the fixed point 0 is superstable and for $\mu \in [4, 6.75]$ the map exhibits its complete period-doubling behavior. While Figure 1.4 shows the magnified version of the original bifurcation (Figure 1.3) in the parameter range $\mu \in [4, 6.75]$ and we study its fixed point, periodic doubling and chaotic behavior. For $4 \leq \mu \leq 5.33$, the fixed point of order one is stable and as the value of parameter μ approaches through 5.33 it bifurcates into periodic points of period 2, 4, 8, and so on, as shown in Figure 1.5. Further, as the value of parameter μ increases through 5.9 a more interesting and complicated behavior, that is, chaos is detected as shown in Figure 1.6, where $5.9 \leq \mu \leq 6.55$. Furthermore, for $\mu > 6.55$ the system again approaches to 0.

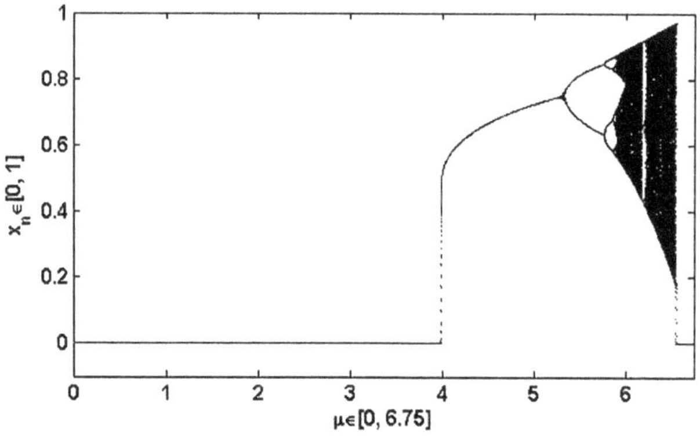

FIGURE 1.3 Period-doubling bifurcation plot for the cubic-type modulated logistic map $M_{2,1}(\mu, x)$ for $0 \leq \mu \leq 6.75$.

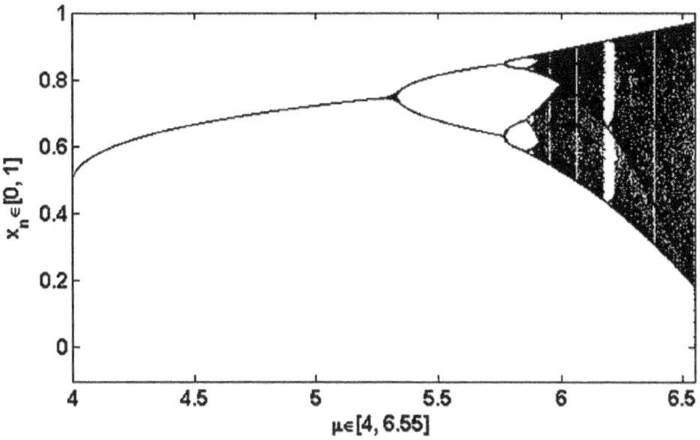

FIGURE 1.4 Period-doubling bifurcation plot for the cubic-type modulated logistic map $M_{2,1}(\mu, x)$ for $4 \leq \mu \leq 6.75/$.

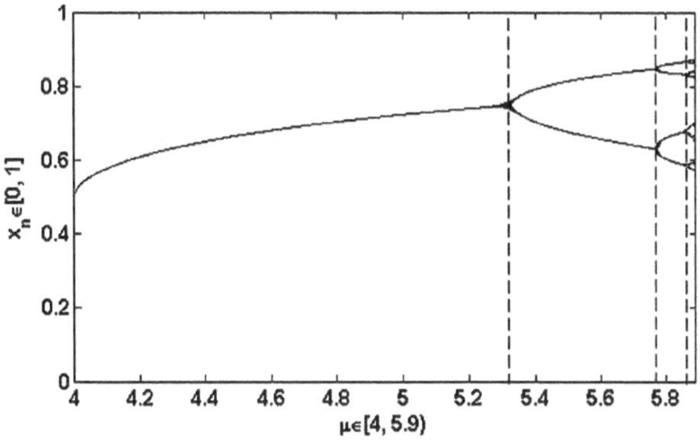

FIGURE 1.5 Period-doubling bifurcation plot for the cubic-type modulated logistic map $M_{2,1}(\mu, x)$ for $4 \leq \mu < 5.9$.

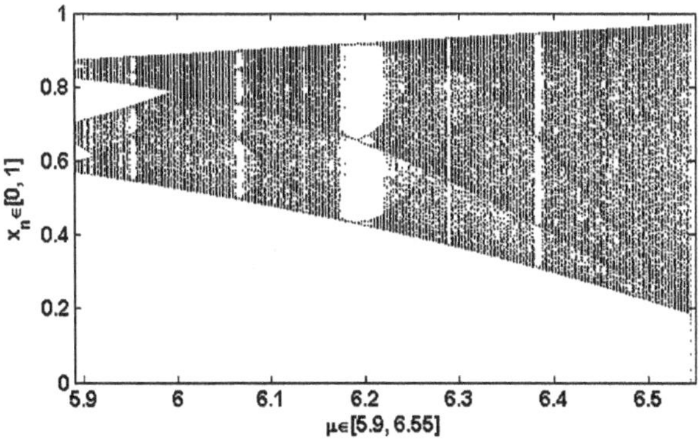

FIGURE 1.6 Period-doubling bifurcation plot for the cubic-type modulated logistic map $M_{2,1}(\mu, x)$ for $5.9 \leq \mu \leq 6.75$.

Remark 1.6

It is interesting for the cubic-type modulated logistic map $M_{2,1}(\mu, x)$ that the fixed point 0 exists in the parameter range $0 \leq \mu < 4$ and $6.55 < \mu \leq 6.75$ and hence it is assumed as super stable fixed point throughout the interval $\mu \in [0, 6.75]$.

1.5 DYNAMICAL ANALYSIS IN $\mu x^p(1 - x)^q$ FOR $p = 1$ AND $q = 2$

In this section, we deal with the dynamics of the modulated logistic map $\mu x^p(1 - x)^q$ for the parameter $p = 1$ and $q = 2$. Therefore, the following cubic-type modulated logistic system is introduced for $x \in [0, 1]$ and $\mu \in [0, 6.75]$:

$$M_{1,2}(\mu, x) = \mu x(1-x)^2 \tag{1.11}$$

Then, from the cubic map $M_{1,2}(\mu, x)$ and Equation (1.3), we determine $x = \frac{1}{3}$ as its unique critical point for which the growth-rate parameter μ attain its maximum and reaches to $\mu_{max} = 6.75$. Therefore, all the dynamical characteristics of the modulated map $M_{1,2}(\mu, x)$ occurs in [0, 6.75] for the parameter μ. Figure 1.3 gives the graphical representation of the family of the parabolic curves of the cubic-type modulated logistic map and shows $x = \frac{1}{3}$ as its critical point for which the functional graph approaches to their maximum value at $\mu = 6.75$. Further, we first study the fixed-point properties of the cubic-type modulated logistic map and then illustrate the dynamics using period-doubling bifurcation.

Theorem 1.7

Let $M_{1,2}(\mu, x) = \mu x(1 - x)^2$ be the cubic-type modulated logistic map defined on [0, 1], for $\mu \in [0, 6.75]$. Then, show that 0 and $1 - \frac{1}{\sqrt{\mu}}$ are the fixed point for the map $M_{1,2}(\mu, x)$, for all $\mu > 0$ and $x \in [0, 1]$.

Proof. Let $M_{1,2}(\mu, x) = \mu x(1 - x)^2$ be the cubic-type modulated logistic map then using Definition 1.1, for the fixed-point x, we have

$$\begin{aligned} &\mu x(1-x)^2 = x, \\ &\mu x(1-x)^2 - x = 0, \\ &x\left(\mu(1-x)^2 - 1\right) = 0, \\ &either, x = 0 \;\; or \;\; x_1 = 1 - \frac{1}{\sqrt{\mu}} \; and \;\; x_2 = 1 + \frac{1}{\sqrt{\mu}} \end{aligned} \tag{1.12}$$

Thus, $x = 0$ and $x_1 = 1 - \frac{1}{\sqrt{\mu}}$ are the fixed points in [0, 1]. Figures 1.7 and 1.8 show the complete functional plot of the fixed points. For $\mu < 1$, the functional plot

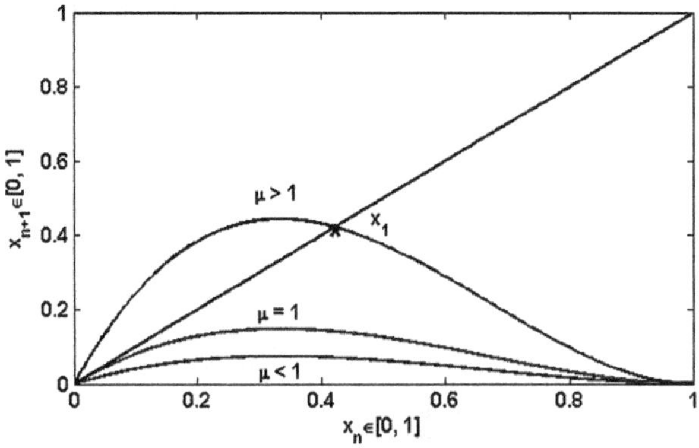

FIGURE 1.7 Functional plot for the cubic-type modulated logistic map $M_{1,2}(\mu, x)$ for $0 \le \mu \le 6.75$.

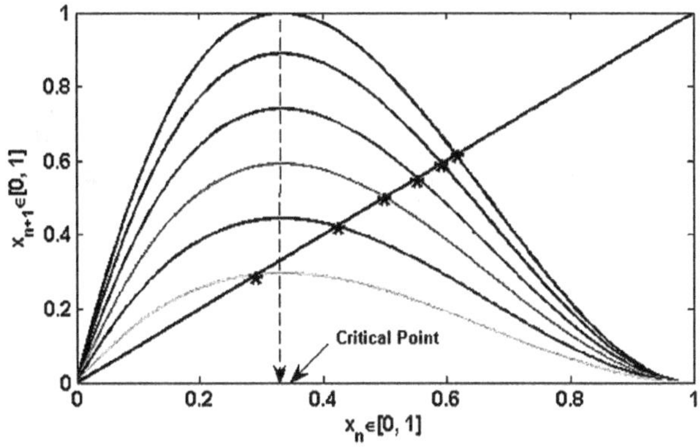

FIGURE 1.8 Functional plot for the cubic-type modulated logistic map $M_{1,2}(\mu, x)$ for $\mu < 1$, $\mu = 1$ and $\mu > 1$.

of the map $M_{1,2}(\mu, x)$ lies below the diagonal $y = x$ that means the origin is the only fixed point in this regime and is stable for $0 \le \mu \le 1$. Also, for $\mu = 1$ the functional plot gets more parabolic making tangent to the diagonal axis and starts to approach the fixed point $1-\frac{1}{\sqrt{\mu}}$. Again, as the value of parameter μ increases through 1, that is, $\mu > 1$ the functional graph intersects the diagonal axis at fixed point $1-\frac{1}{\sqrt{\mu}}$. and is always stable for $1 < \mu < 4$ as shown in Figure 1.8.

Theorem 1.8

Let $M_{1,2}(\mu, x) = \mu x(1 - x)^2$ be the cubic-type modulated logistic map defined on [0, 1], for $\mu \in [0, 6.75]$. Then, show that $M_{1,2}(\mu, 0) = 0 = M_{1,2}(\mu, 1)$.

Proof. Let us consider $M_{1,2}(\mu, x) = \mu x(1 - x)^2$ be the cubic-type modulated logistic map, then substituting $x = 0$, we get $M_{1,2}(\mu, x) = 0$ and also taking $x = 1$, we have $M_{1,2}(\mu, 1) = 0$. Figures 1.7 and 1.8, shows the graphical representation for $M_{1,2}(\mu, 0) = 0 = M_{1,2}(\mu, 1)$.This completes the proof.

Theorem 1.9

Let $M_{1,2}(\mu, x)$ be the modulated map defined on [0, 1]. Then, the fixed-point $x = 0$ is stable when $0 \le \mu \le 1$ and $1-\frac{1}{\sqrt{\mu}}$ is stable when $1 < \mu < 4$ for the map $M_{1,2}(\mu, x)$.

Proof. By using Definition 1.2 for the stabilization of fixed points we have to prove that $\left|M_{1,2}'(\mu,x)\right| < 1$. Thus, let us consider

$$\left|M_{2,1}'(\mu,x)\right| = \left|\mu(1-x)^2 - 2\mu x(1-x)\right| \tag{1.13}$$

Substituting the fixed-point $x = 0$ in (1.13), then clearly $\left|M_{2,1}'(\mu,x)\right| < 1$, that means, the fixed point 0 is stable for $0 \le \mu \le 1$. Now, putting $1-\frac{1}{\sqrt{\mu}}$ in (1.13) then, we have

$$\left|M_{2,1}'\left(\mu,1-\frac{1}{\sqrt{\mu}}\right)\right| = \left|\mu\left(1-1-\frac{1}{\sqrt{\mu}}\right)^2 - 2\mu\left(1-\frac{1}{\sqrt{\mu}}\right)\left(1-1-\frac{1}{\sqrt{\mu}}\right)\right|,$$

$$\left|M_{2,1}'\left(\mu,1-\frac{1}{\sqrt{\mu}}\right)\right| = \left|\mu\left(\frac{1}{\sqrt{\mu}}\right)^2 - 2\mu\left(1-\frac{1}{\sqrt{\mu}}\right)\frac{1}{\sqrt{\mu}}\right|,$$

$$\left|M_{2,1}'\left(\mu,1-\frac{1}{\sqrt{\mu}}\right)\right| = \left|1-2\sqrt{\mu}\left(1-\frac{1}{\sqrt{\mu}}\right)\right|,$$

$$\left|M_{2,1}'\left(\mu,1-\frac{1}{\sqrt{\mu}}\right)\right| = \left|1-2\left(\sqrt{\mu}-1\right)\right|,$$

$$\textit{that is,} \left|M_{2,1}'\left(\mu,1-\frac{1}{\sqrt{\mu}}\right)\right| < 1,$$

for each $1 < \mu < 4$. Hence, the fixed point $1-\frac{1}{\sqrt{\mu}}$ is stable. This completes the proof.

Example 1.10

Let $\mu x(1 - x)^2$ be the cubic-type modulated logistic map on [0, 1] for $\mu \in [0, 6.75]$. Then, the fixed point 0 and 0.4226 are stable at $\mu = 0.75$ and 3, respectively.

Solution

Using Definition 1.2, for the stability of fixed points 0 and 0.4226 we will show that $\left|M'_{1,2}(\mu,x)\right| < 1$. Therefore, we consider

$$\left|M'_{2,1}(\mu,x)\right| = \left|\mu(1-x)^2 - 2\mu x(1-x)\right|, \tag{1.14}$$

where $x \in [0, 1]$ and $\mu \in [0, 6.75]$. Then, substituting $x = 0$ and $\mu = 0.75$ in the relation (1.14) we get $\left|M'_{1,2}(0.75,0)\right| < 1$. Hence the fixed point 0 is stable. Now, again putting $x = 0.4226$ and $\mu = 3$ in relation (1.14), we get

$$\left|M'_{1,2}(3,0.4226)\right| = \left|3\times(1-0.4226)^2 - 2\times3\times0.4226\times(1-0.4226)\right| = \left|1-1.4645\right|,$$

that is, $\left|M'_{1,2}(3,0.4226)\right| = \left|-0.4645\right| < 1.$

Thus, the fixed-point $x = 0.4226$ is also stable for $\mu = 3$.

Remark 1.11

From Sections 1.4 and 1.5 it is observed that for the control parameter $(p, q) = (2, 1)$ and $(1, 2)$ the modulated logistic map $M_{p,q}(\mu, x)$ has same growth-rate parameter range, that is, $\mu \in [0, 6.75]$.

1.5.1 Period-Doubling Bifurcation Analysis for the Map $M_{1,2}(\mu, x)$

For $p = 1$ and $q = 2$, the modulated logistic map $\mu x^p(1 - x)^q$ reduces into the cubic-type modulated system $\mu x(1 - x)^2$, where $x \in [0, 1]$ and $\mu \in [0, 6.75]$. In Figure 1.9 the period-doubling bifurcation is plotted for $\mu \in [0, 6.75]$ and the dynamical characteristics from period-doubling to chaos are illustrated for the cubic system $M_{1,2}(\mu, x)$. Further, for $1 < \mu < 4$, the trajectory shows the stability of the fixed point $1 - \frac{1}{\sqrt{\mu}}$ as shown in Figure 1.10 and as the parameter μ approaches through 4 the first period-doubling occurs and the trajectory bifurcates into periodic points of period-2. But for $\mu > 5$, the trajectory vibrates into higher-order periodic points of period 2^n, $n \in N$ and it continues up to 5.33. Furthermore, it is remarkable to see that as μ approaches to 5.33 the system starts to become more sensitive and the beauty of chaos is detected in the parameter range $5.33 < \mu \le 6.75$, as shown in Figure 1.11. But Figure 1.12 shows that as the parameter move in the range $6.01 \le \mu \le 6.12$ the system again turn back into the stable period-doubling of order-3.

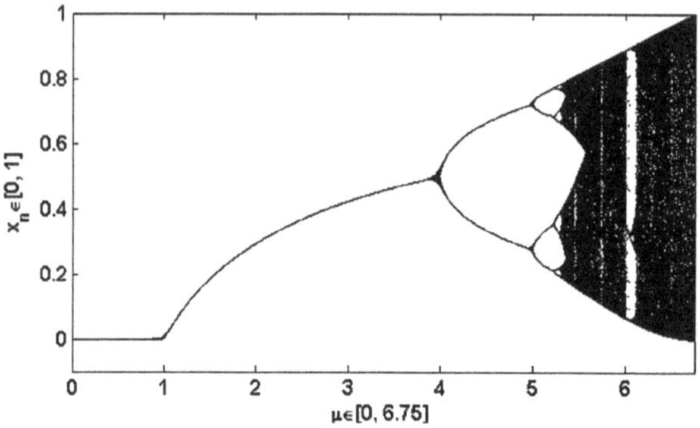

FIGURE 1.9 Period-doubling bifurcation plot for the cubic-type modulated logistic map $M_{1,2}(\mu, x)$ for $0 \leq \mu \leq 6.75$.

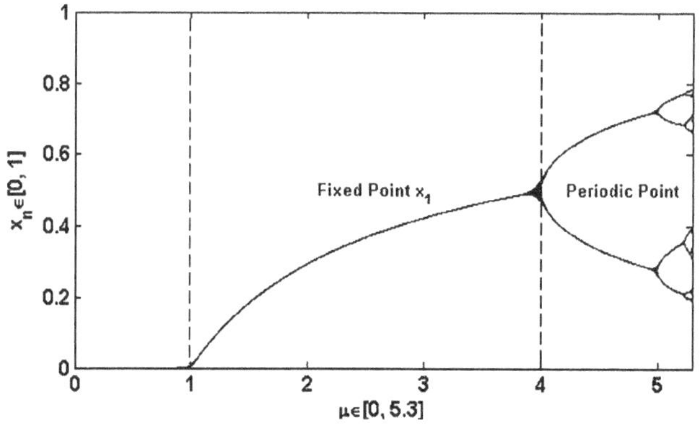

FIGURE 1.10 Period-doubling bifurcation plot for the cubic-type modulated logistic map $M_{1,2}(\mu, x)$ for $0 \leq \mu \leq 5.33$.

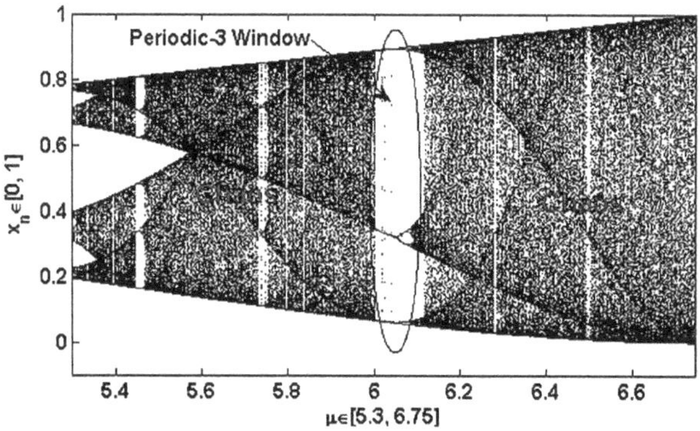

FIGURE 1.11 Period-doubling bifurcation plot for the cubic-type modulated logistic map $M_{1,2}(\mu, x)$ for $5.33 < \mu \leq 6.75$.

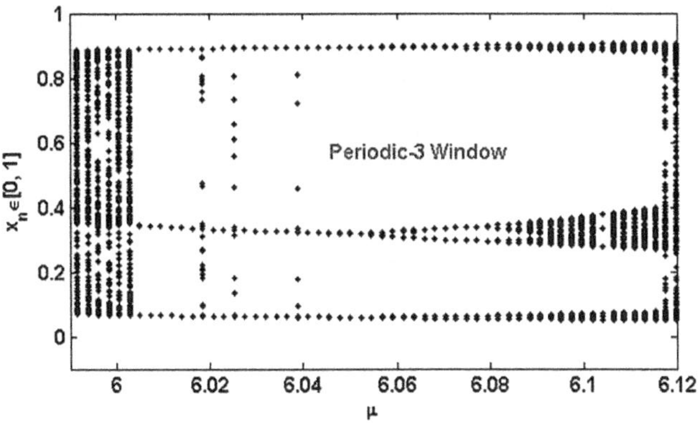

FIGURE 1.12 Period-doubling bifurcation plot for the cubic-type modulated logistic map $M_{1,2}(\mu, x)$ for $6.01 \le \mu \le 6.12$.

1.6 DYNAMICAL ANALYSIS IN $\mu x^p(1 - x)^q$ FOR $p = 2$ AND $q = 2$

As studied in the earlier sections, in this section for $p = q = 2$ we deal with the new quartic-type modulated logistic map $M_{2,2}(\mu, x) = \mu x^2(1 - x)^2$. The critical point of the map $M_{2,2}(\mu, x)$ using Equation (1.3) is calculated as $M_{2,2}(\mu, x)$,which is also unique. Also, the maximum of the growth-rate parameter μ approaches to $\mu_{max} = 16$, which means the growth-rate parameter varies between 0 and 16. The following analytical results are obtained from the study:

Theorem 1.12
Let $M_{2,2}(\mu, x) = \mu x^2(1 - x)^2$ be the quartic-type modulated logistic map on [0, 1] and $\mu \in [0, 16]$. Then, determine the fixed point for the map $M_{2,2}(\mu, x)$, for all $x \in [0, 1]$.

Proof. Let $M_{2,2}(\mu, x) = \mu x^2(1 - x)^2$ be the quartic-type modulated logistic map then using Definition 1.1, for the fixed point, we can write

$$\mu x^2(1-x)^2 = x,$$

$$x\left(\mu x(1-x)^2 - 1\right) = 0,$$

either, $x = 0$

or

$$x_1 = \frac{1}{3}\left(2 + \frac{2^{\frac{1}{3}}\mu}{\left(27\mu^2 - 2\mu^3 + 3\sqrt{3}\sqrt{27\mu^4 - 4\mu^5}\right)^{\frac{1}{3}}} + \frac{\left(27\mu^2 - 2\mu^3 + 3\sqrt{3}\sqrt{27\mu^4 - 4\mu^5}\right)^{\frac{1}{3}}}{2^{\frac{1}{3}}\mu}\right).$$

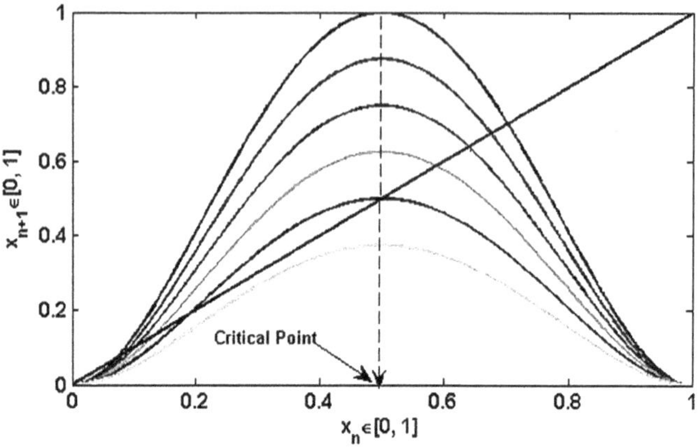

FIGURE 1.13 Functional plot for the quartic-type modulated logistic map $M_{2,2}(\mu, x)$ for $0 \leq \mu \leq 16$.

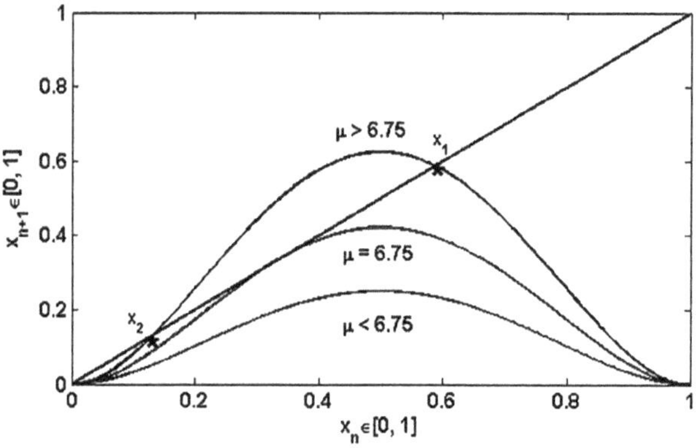

FIGURE 1.14 Functional plot for the quartic-type modulated logistic map $M_{2,2}(\mu, x)$ for $\mu < 6.75$, $\mu = 6.75$ and $\mu > 6.75$.

Thus, the points x and x_1 are the required real and stable fixed points for the map $M_{2,2}(\mu, x)$. Figure 1.13 shows the complete functional plot for the quartic-type modulated logistic map and Figure 1.14 represents that for $\mu < 6.75$ there exist only 0 fixed point, for $\mu = 6.75$ the functional plot touches the diagonal axis and for $\mu > 6.75$ the functional graph intersects the diagonal axis at the real stable fixed point x_1.

Theorem 1.13

Let $M_{2,2}(\mu, x) = \mu x^2(1 - x)^2$ be the modulated logistic map defined on [0, 1] and $\mu \in [0, 16]$. Then, show that $M_{2,2}(\mu, 0) = 0 = M_{2,2}(\mu, 1)$.

Proof. Similar to the proof of Theorems 1.2 and 1.8

Remark 1.14

Similar to Theorems 1.3 and 1.9, it is analysed that for the map $\mu x^2(1-x)^2$ on [0, 1], where $\mu \in [0, 16]$, the fixed-point $x = 0$ is stable for $0 \le \mu \le 16$, that is, it is super stable throughout the interval and x_1 is stable for $6,8 < \mu < 10.5$.

Example 1.15

Let $\mu x^2(1-x)^2$ be the quartic-type logistic map defined on [0, 1], where $\mu \in [0, 16]$. Then, determine the stability of the fixed points 0 and 0.5509 for $\mu = 5$ and $\mu = 9$, respectively.

Solution

Using Definition 1.2, for the stability of fixed points we will show that $\left|M'_{2,2}(\mu,x)\right| < 1$. Then, for $x \in [0, 1]$ and $\mu \in [0, 16]$, we have

$$\left|M'_{2,2}(\mu,x)\right| = \left|2\mu x(1-x)^2 - 2\mu x^2(1-x)\right|. \quad (1.15)$$

Therefore, substituting $x = 0$ and $\mu = 5$ in relation (1.15) we obtain $\left|M'_{2,2}(5,0)\right| < 1$. Hence the fixed point 0 is stable. Again, putting $x = 0.5509$ and $\mu = 9$ in (1.15), we have

$$\left|M'_{2,2}(9,0.5509)\right| = \left|2\times 9\times 0.5509(1-0.5509)^2 - 18\times(0.5509)^2(1-0.5509)\right|,$$

$$= \left|9.9162\times(1-0.5509)^2 - 5.463\times(1-0.5509)\right|,$$

$$= \left|9.9162\times(0.4491)^2 - 5.463\times 0.4491\right|,$$

$$= \left|2 - 2.4534\right|,$$

$$\left|M'_{2,2}(9,0.5509)\right| = \left|-0.4534\right| < 1.$$

Thus, the fixed point $x_1 = 0.5509$ is also stable for $\mu = 9$.

1.6.1 Period-Doubling Bifurcation Analysis for the Map $M_{2,2}(\mu, x)$

For $p = 2$ and $q = 2$, the Figures 1.15–1.18 give a complete graphical representation of the dynamics of quartic-type modulated logistic map $\mu x^2(1-x)^2$. Taking the step size of $\mu = 0.001$, starter $x_0 = 0.7$ and the number of iterations $N = 300$ the period-doubling is plotted and the properties from period-doubling to irregular behavior are studied. For $6.8 < \mu < 10.5$, the fixed point x_1 is always stable, as given in Remark 1.14. When $\mu > 10.5$, the iterative orbit for the quartic map bifurcates into two periodic points of period-2, which is also the first bifurcation of it as shown in Figure 1.17. Proceeding in this way as μ approaches through 11.6 the period-doubling bifurcation continues

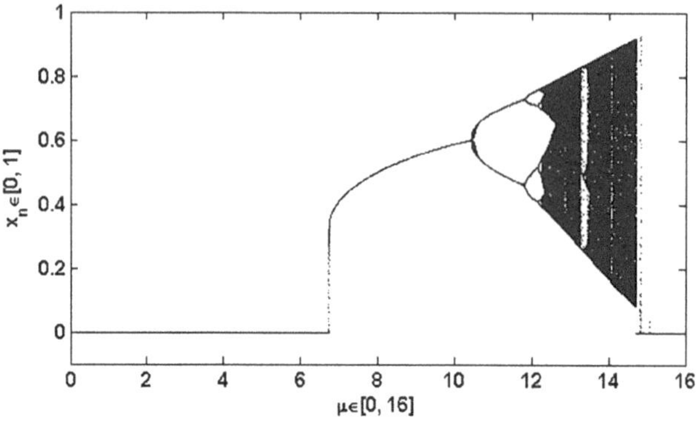

FIGURE 1.15 Period-doubling bifurcation plot for the quartic-type modulated logistic map $M_{2,2}(\mu, x)$ for $0 \leq \mu \leq 16$.

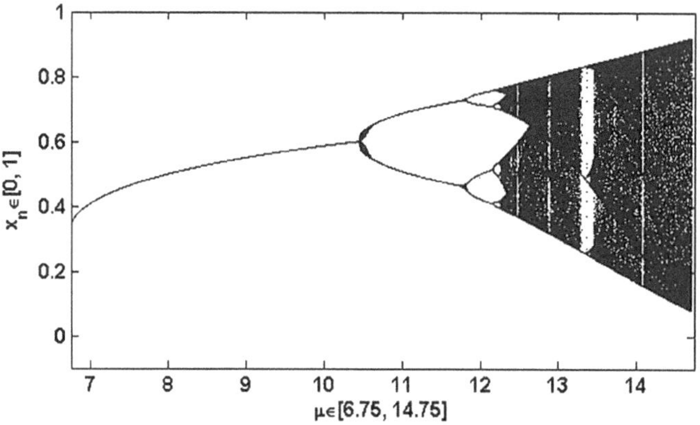

FIGURE 1.16 Period-doubling bifurcation plot for the quartic-type modulated logistic map $M_{2,2}(\mu, x)$ for $6.75 \leq \mu \leq 14.75$.

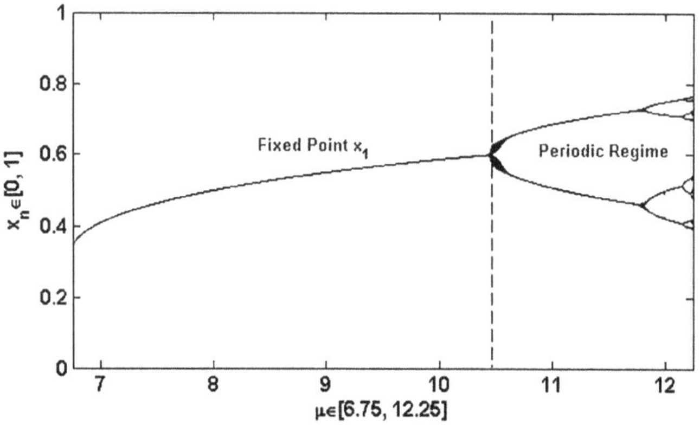

FIGURE 1.17 Period-doubling bifurcation plot for the quartic-type modulated logistic map $M_{2,2}(\mu, x)$ for $6.75 < \mu \leq 12.25$.

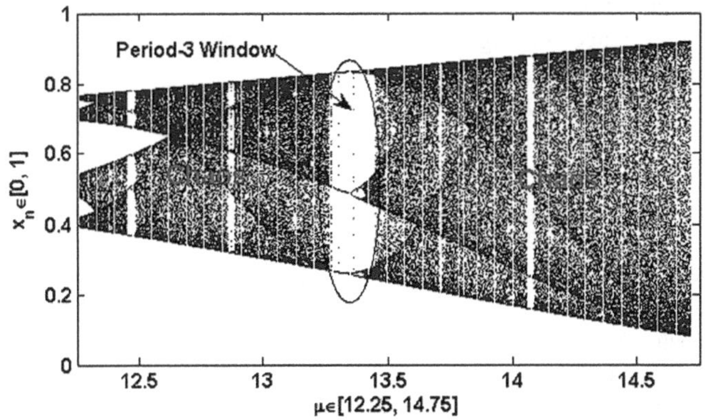

FIGURE 1.18 Period-doubling bifurcation plot for the quartic-type modulated logistic map $M_{2,2}(\mu, x)$ for $12.25 \leq \mu \leq 14.75$.

till 12.2. Further, it is amazing to see that, as $\mu \approx 12.2$, the period-doubling starts to approach an irregular behavior; that is, chaos occurs as shown in Figure 1.18. Furthermore, it is illustrated that due to the super stability behavior of fixed point 0 it again undergoes a stability behavior for $14.7 < \mu \leq 16$.

1.7 CONCLUSION

Throughout this chapter, we study the dynamical nature of the modulated logistic map $\mu x^p(1 - x)^q$, where $p, q > 0$, and examine its characteristics for the parameter values $(p, q) = (2, 1), (1, 2)$ and $(2, 2)$. The analytical as well as numerical simulations are established for the cubic- and quartic-type modulated logistic maps. Further, it is analysed that the modulated logistic map for the parameter values $(p, q) = (2, 1)$ and $(1, 2)$ have the similar range of interval for the growth-rate parameter μ. Although, the subintervals for both type of maps yields different dynamical properties.

In Section 1.3 the basic preliminary results in the dynamics of modulated logistic map are established and then in Section 1.4 the dynamics of the cubic type modulated logistic map $\mu x^2(1 - x)$ is studied for the growth-rate parameter $\mu \in [0, 6.75]$. The critical point, maximum of growth-rate parameter μ, fixed and periodic point and the chaotic behavior are established followed by some theorems, remarks and examples. In addition, all the results are shown by graphical representations using period-doubling bifurcation plot.

In Section 1.5, another cubic-type modulated logistic map $\mu x(1 - x)^2$ is introduced for the growth-rate parameter $\mu \in [0, 6.75]$ and all the dynamical properties are examined using mathematical analysis. The numerical simulation is presented using period-doubling bifurcation in Figures 1.9–1.12.

While in Section 1.6, a quartic type modulated logistic system $\mu x^2(1 - x)^2$, where $\mu \in [0, 16]$ is found and the dynamical properties are examined followed by graphical simulation, as shown in Figures 1.15–1.18. Therefore, the proposed modulated logistic map enriches the dynamical characteristics with an extra degree of freedom in the parameters p and q, which means the map can be used for the modeling of various applications in science and engineering, such as communication security, traffic control model, cryptography, etc.

REFERENCES

1. May R., Simple mathematical models with very complicated dynamics. *Nature*, 261 (1976), pp. 459–475.
2. May R. and Oster G. F., Bifurcations and dynamic complexity in simple ecological models. *Amer. Nat.*, 101 (1976), pp. 573–599.
3. Feigenbaum M. J., Quantitative universality for a class of nonlinear transformations. *J. Stat. Phys.*, 19 (1) (1978), pp. 25–52.
4. Devaney R. L., *An Introduction to Chaotic Dynamical Systems*, CRC Press, New York, (1948).
5. Devaney R. L., *A First Course in Chaotic Dynamical Systems: Theory and Experiment*, CRC Press, New York, (1992).
6. Holmgren R. A., *A First Course in Discrete Dynamical Systems*, Springer-Verlag, New York, (1994).
7. Block L. S. and Coppel W. A., *Dynamics in One Dimension*, Springer-Verlag, New York, (1992).
8. Wiggins S., *Introduction to Applied Nonlinear Dynamics and Chaos*, Springer-Verlag, New York, (1990).
9. Alligood K. T., Sauer T. D. and Yorke J. A., *Chaos: An Introduction to Dynamical Systems*, Springer-Verlag, New York, (1996).
10. Ausloos M. and Dirickx M., *The Logistic Map and the Route to Chaos: from the Beginnings to Modern Applications*, Springer-Verlag, New York, (2006).
11. Elagdi S. N., *Chaos: An Introduction to Difference Equations*, Springer-Verlag, New York, (1999).
12. Martelli M., *Chaos: An Introduction to Discrete Dynamical Systems and Chaos*, Wiley-Interscience, New York, (1999).
13. Strogatz S. H., *Nonlinear Dynamics and Chaos*, Persus Books Publishing, New York, (1994).
14. Smith, J. M., *Mathematical Ideas in Biology*, Cambridge University Press, Cambridge, (1968).
15. Briden W. and Zhang S., Stability of solutions of generalized logistic difference equations. *Period. Math. Hung.*, 9 (1994), pp. 81–87.
16. Willeboordse, F. H., The spatial logistic map as a simple prototype for spatiotemporal chaos. *Chaos*, 13 (2003), pp. 533–540.
17. Baptista M. S. and Caldas I. L., Dynamics of the kicked logistic map. *Chaos Solit. Fract.*, 7 (1976), pp. 325–336.
18. Nandi A., Dutta D., Bhattacharjee J. K. and Ramaswamy R., The phase-modulated logistic map. *Chaos*, 15 (2005), 023107.
19. Elhadj Z. and Sprott J. C., The effect of modulating a parameter in the logistic map. *Chaos*, 18 (2008), 023119.
20. Wu G. C. and Baleanu D., Discrete fractional logistic map and its chaos. *Nonlin. Dyn.*, 75 (2014), pp. 283–287.
21. Radwan A. G., On some generalised discrete logistic map. *J. Adv. Res.*, 4 (2013), pp. 163–171.

22. Sayed S. W., Radwan A. G. and Fahmy A. H., Design of positive, negative and alternating sign generalized logistic maps. *Discrete Dyn. Nat. Soc.*, (2015), 586783, 23pages.
23. Ashish, Cao J. and Chugh R., Chaotic behavior of logistic map in superior orbit and an improved chaos-based traffic control model. *Nonlinear Dyn.*, 94 (2) (2018), pp. 959–975.
24. Ashish, Cao J., A novel fixed-point feedback approach studying the dynamical behavior of standard logistic map. *Int. J. Bifurcat. Chaos*, 29 (1) (2019), 1950010 (16 pages).
25. Ashish, Cao J. and Chugh R., Controlling chaos using superior feedback technique with applications in discrete traffic models. *Int. J. Fuzzy Syst.*, 21 (2019), pp. 1467–1479.
26. Rani M. and Agarwal R., A new experimental approach to study the stability of logistic map. *Chaos Solit. Fract.*, 41 (2009), pp. 2062–2066.
27. Kanso A. and Smaoui N., Logistic chaotic maps for binary numbers generations. *Chaos Solit. Fract.*, 40 (2009), pp. 2557–2568.
28. Andrecut M., Logistic map as a random number generator. *Int. J. Mod. Phys. B.*, 12 (921) (1998), pp. 101–102.
29. Pareek N. K., Patidar V. and Sud K. K., Image encryption using chaotic logistic map. *Imag. Vis. Comput.*, 24 (2006), 926–934.
30. Baptista M. S., Cryptography with chaos. *Phys. Lett. A*, 240 (1998), pp. 50–54.
31. Hafiz S. K., Radwan A. G. and Haleem S. H., Encryption applications of a generalized chaotic map. *Appl. Math. Inf. Sci.*, 9 (6) (2015), pp. 1–19.
32. Wang X. and Xu D., Image encryption using genetic operators and intertwining logistic map. *Nonlin. Dyn.*, 7 (2014), pp. 2975–2984.
33. Singh N. and Sinha A., Chaos-based secure communication system using logistic map. *Opt. Lasers Eng.*, 48 (2010), pp. 398–404.
34. Salarieh H. and Alasty A., Chaos control in uncertain dynamical systems using nonlinear delayed feedback. *Chaos Solitons Fractals*, 41 (2009), pp. 67–71.
35. Medina A. P., Rio-Correa J. L., and Hernandez J. L., Design of chaotic analog noise generators with logistic map and MOS QT circuit. *Chaos Solit. Fract.*, 40 (4) (2009), pp. 1779–1793.
36. Diamond P., Chaotic behaviour of systems of difference equations. *Int. J. Systems Sci.*, 7 (8) (1976), pp. 953–956.
37. Robinson C., *Dynamical Systems: Stability, Symbolic Dynamics, and Chaos*, CRC Press, Florida, (1995).
38. Sharkovsky A. N., Maistrenko Yu. L. and Romanenko E. Yu., *Difference Equations and Their Applications*, Kluwer Academic Publisher, Netherlands, (1993).

2 A Survey on Evolutionary Clustering Algorithms and Applications

Chandra Mani Sharma and Shail Kr Dinkar

CONTENTS

2.1 INTRODUCTION

Evolutionary approaches form a subset of modern artificial algorithms, which are inspired by nature. Evolutionary approaches are an efficient and effective choice in order to solve complex, continuous and discrete, convex and nonconvex, nonlinear and linear, optimization problems [1–4]. The evolutionary approaches are inspired from the Darwinian theory of evolution developed by the English naturalist Charles Darwin in the nineteenth century, which stated that the natural evolutionary survival of species takes place through natural selection, which acts so as to increase an individual's ability to survive, excel and reproduce. This theory includes the broad concepts of transmutation of species and evolution.

DOI: 10.1201/9781003089636-2

2.1.1 What Are the Evolutionary Algorithms?

Evolutionary algorithms are also known as metaheuristic optimization techniques. They have operations such as selection, mutation, reproduction, etc., all of which help to find out the best solutions to optimization problems.

Figure 2.1 shows the typical five steps of an evolutionary approach for optimization. These steps are described as follows:

Step 1 – Random Initialization of Population: In this step, the solutions to an optimization problem are randomly assigned. This corpus of the solutions is also known as the population. These are the candidate solutions, which are weak in nature. The objective is to improve the viability of these solutions by generating new ones.

Step 2 – Population Fitness Evaluation: In the second step, each individual (candidate solution in corpus) is assessed based on a fitness criterion defined by fitness function. Fitness function gives the measure of how fit a solution is in the given population. It is a relative measure and helps establish the superiority of some solutions. In other words, fitness function is the objective function for the given optimization problem. There are two types of fitness functions: fixed and mutable. The fixed fitness function does not change over the course of the algorithm execution, whereas the mutable fitness function can change at some or the other point of time while executing the evolutionary algorithm.

Step 3 – Breed New Solutions: This is the most important step of any evolutionary algorithm, where new candidates are generated. There exist many ways of generating new candidate solutions, and each approach has some predefined operation for its achievement. The genetic algorithm has operators, such as mutation and crossover, to generate new candidate solutions. Mutation is

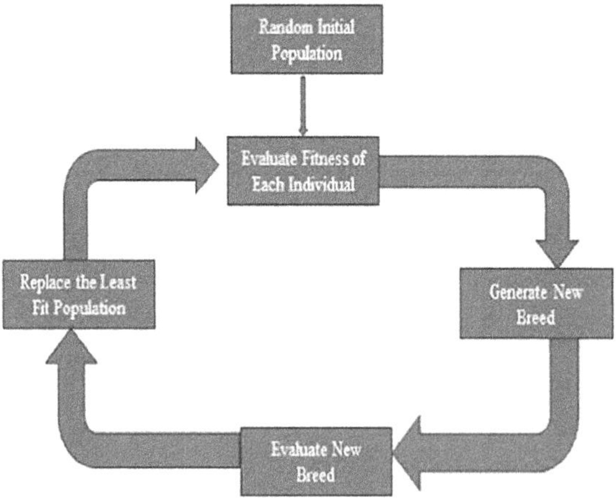

FIGURE 2.1 Steps in a typical evolutionary approach for optimization.

unary operator and crossover is the binary operator. With the help of mutation, we can mutate a given solution to create a new one. On the other hand, crossover requires two solutions and generates a third solution as the offspring.

Step 4 – Evaluate the New Breed: Each newly created solution is tested for fitness on the given criterion using the fitness function. If it is good enough, then it becomes part of the population.

Step 5 – Replace the Least Fit Solution: After passing the fitness criterion, new solutions replace the already existing solutions in the population. Strong offspring replace the weak members of the population.

The above steps are to be repeated unless the best solution is achieved or a finite number of iterations have passed.

2.1.2 Evolutionary Algorithms for Solving Optimization Problems

There are many evolutionary approaches to solve various optimization problems, including clustering [5–8]. The list of some of them is given as follows:

Genetic algorithm [9]
Differential evolution [10]
Particle swarm optimization [11]
Ant colony optimization [12]
Artificial bee colony [13]
Grey Wolf optimizer [14]
Whale optimizer [15]
Glowworm swarm optimization [16]
Gravitational search algorithm [17, 18]
Central force optimization [19]
Harmony search optimization [20]
Cuckoo search [21]
Bat algorithm [22]
Antlion optimizer [23]

2.1.3 Clustering Approaches and Their Categorical Division

Clustering is the approach of dividing a given data sets into a finite number of groups, which are known as data clusters. The clusters are created in such as a way that similar data points fall in the same cluster and dissimilar data points fall in different clusters. This is a long-standing problem and researchers have been trying to solve it in different ways. In machine learning, clustering can be done using unsupervised learning, wherein based on some similarity measure, groups of data are formed. There exist many clustering approaches, like k-means, c-means (a fuzzy version of k-means), mean shift clustering, DBSCAN, agglomerative clustering, expectation maximization clustering, evolutionary clustering etc.

Based on the approach used for clustering, the methods can be divided into following five categories:

Centroid-based-clustering methods: In this type of approaches, we decide the number of clusters K beforehand. Different clusters are formed in such a way that the distance of the data points from centroid is minimized. It turns out to be a NP-hard problem. (NP-hard refers to a set of non-deterministic polynomial-time problems. A subset sum problem is an example of NP-hard problem.) Therefore, the solutions are approximated to a fixed number of trials. These methods are iterative in nature and try to improve over each iteration. K-means clustering is an example of a centroid-based clustering method. Centroid-based approaches for clustering have been very popular for this task, but their performance depends largely on the initial cluster centers and these approaches happen to converge to the local minima [24].

Density-based-clustering methods: In density-based clustering methods, the clusters are formed by grouping the points together, which share the common scattering density properties. These methods are somewhat similar to the centroid-based methods. The typical examples of density-based methods for data clustering include DBSACN (**Density-based spatial clustering of applications with noise**) and OPTICS (**Ordering points to identify cluster structure**). In literature, there are variants and improvements available for density-based clustering, as suggested in [25–28]. In the work of He et al. [25], they proposed an interesting version of DBSCAN called MR-DBSCAN, which uses parallel computing and MapReduce for forming clusters. Kisilevich et al. in [26] describe an approach called P-DBSCAN – a density-based approach for geo-spatial data. Both Lv et al. [27] and Kumar et al. [28] proposed the variants of DBSCAN.

Distribution-based-clustering methods: These methods follow some probability distribution for assigning a point into a cluster. The probability distribution may be normal (Gaussian) distribution, binomial distribution, chi square distribution or some other distribution. Normal distribution is more prevalent in this type of clustering. EM-based clustering is an example that uses this approach for grouping the data into different clusters. In a recent improvement for distribution-based clustering, Wang at al. [29] proposed a local distribution-based technique for clustering. This technique overcomes the problem of existing Bayesian approaches for clustering. The limitation of Bayesian clustering approaches is their inability to tackle the imbalanced data. The technique proposed in [29] works well with imbalanced data. In another work, Güngör et al. [30] proposed an approach to find clusters in data using Gaussian Distributed Density (GDD) that exploits the density and distance metrics of the given data sample.

Hierarchy-based-clustering methods: These methods of clustering take into account the proximity of data points to form clusters. They assume that two points, which are near to each other, share more of the similarity as compared to those situated at a distance. In this way a hierarchy of clustering points is formed. That is why these approaches are also called as connectivity-based clustering methods. Scalability is one major issue with these methods. They fall under two subcategories, namely agglomerative hierarchical clustering methods and divisive hierarchical clustering methods. The real world data grows exponentially and it is harder to keep pace with it for the purposes of analysis. For this situation, hierarchical algorithms come in handy, as the growth in data is hierarchical. Researchers have been working in this field and many advanced algorithms have been proposed in the literature. Bouguettaya et al. [31] proposed an agglomerative hierarchical approach for clustering that is capable of handling the almost exponentially increasing data. A survey on hierarchical clustering algorithms and applications can be found in [32].

Subspace clustering methods: At present, the high dimension of data is a challenge for data mining algorithms. (Typically, a data set is called high dimensional when number of features turns out to be bigger than number of observations in the data set.) In unsupervised machine learning also, the high dimension of data poses many challenges. Subspace clustering algorithms make clusters of the data points falling into low dimensional points in a given data space. Subspace clustering approaches work in two ways – either top-down or bottom-up. A clique algorithm is an example of a subspace clustering algorithm. There exist various versions of subspace clustering algorithms. Working accurately with noisy data is a prevalent issue with subspace clustering methods. These methods are more suitable for computer vision-based applications. Wang et al. [33] proposed a subspace clustering technique that works well with noisy data. Cao et al. [34] described another subspace clustering for computer vision application.

2.2 LITERATURE SURVEY AND RELATED WORK

Berkhin et al. [4] provide a good survey of the traditional clustering techniques used in data mining. The traditional algorithms need to be scaled up to meet the requirements of modern-day applications. This is an era of big data analytics and, often, clustering is required to be performed on big data [5–7]. A survey of various clustering methods, for massive data, has been presented by Guyeux et al. [5]. With the advancement in computational infrastructure and improvements in methodologies, it is now possible to easily process huge amounts of data. Parallel processing and algorithms like MapReduce help achieve this objective. In Aljarah et al. [6], they propose an approach exploiting the power of parallel processing and MapReduce and deploying particle swarm optimization (PSO) for the task of clustering. This approach performs well on synthetic as well as real data and overcomes some of problems of the traditional PSO approach for clustering.

In another work by Al-Madi et al. [7], a parallel MapReduced based clustering method has been proposed. The method uses glowworm swarm optimization to form clusters of data with the efficient use of parallel processing and the MapReduce.

They formulate the clustering task as a multi-modal optimization problem to find the best centroids for clustering segments.

There are many hybrid variants of popular evolutionary approaches, such as PSO. In [8], a hybrid approach using PSO and simulated annealing (SA) has been described. It is a general practice with modern applications to optimize multiple objective functions simultaneously.

In [35], an improved version, namely Opposition based antlion optimizer using Cauchy distribution to solve clustering problem, was outlined. This modified version utilizes K-means clustering mechanism as objective function to minimize the intra-cluster variance. The performance of proposed algorithm is evaluated over six datasets taken from the UCI machine learning repository and comparison is performed with classical ALO and a recently developed version namely Opposition-based ALO using Laplace distribution.

These methods can be classified based on the different objective functions, encoding strategies used, evolutionary operators, approach for the selection of the terminal solution, and the mechanism to maintain non-dominated solutions [36]. Researchers have been trying to solve the task of multi-objective clustering using evolutionary approaches. Various techniques have been proposed with regard to this issue. A survey on multi-objective clustering techniques can be found in [37]. In some application areas, the growth of data is incremental, meaning that we need to keep track of the clusters at different time steps. The static clustering approaches (where data are fixed in dimension and size) may not work in this scenario. Adaptive clustering is a good solution for such scenarios. In literature, there exist adaptive evolutionary strategies for clustering. These techniques are good for many applications where data are dynamic. Xu et al. [37] present a framework for adaptive evolutionary clustering. They used an adaptive forgetting factor in the proposed technique, minimizing the mean squared error for tracking the incremental changes. This approach was used to find out clusters on real and synthetic data. Clustering can be solved as a combinatorial optimization problem. The qualitative outcome of a given nature-inspired optimization method depends on the right combination of operators being used for exploration and evolution tasks. The weakness of this combination takes the algorithm to premature convergence, sticking at local optima and making it unable to reach the global optima [38]. Dinkar et al. [38] proposed an opposition-based Laplacian ant lion optimizer that outperforms the traditional Laplacian ant lion optimizer. In their another work, Dinkar et al. [39] proposed a Lévy Flight Antlion optimizer based on the idea of opposition for the solving a range of optimization problems. There exist many other nature-inspired and evolutionary algorithms, such as artificial bee colony optimization (ABC), big-bang big-crunch algorithm, continuous scatter search (CSS), cuckoo search, differential evolution, evolutionary programming, evolutionary programming, firefly algorithm, gene expression programming, genetic algorithm, genetic programming, gravitational search algorithm, harmony search, league championship algorithm, neuro-evolution, population-based incremental learning (PBIL), particle swarm optimization (PSO), etc.

2.3 APPLICATIONS OF EVOLUTIONARY CLUSTERING ALGORITHMS

There are many applications of evolutionary clustering algorithms [3, 36, 40–53]. Depending on the nature of data and category of domain, they work reliably in order to find the clusters efficiently and effectively. These approaches have applications in text document clustering, image segmentation, customer segmentation, telecommunication clustering, healthcare analysis, pollution effect analysis etc. The key advantages of evolutionary algorithms include the flexibility and self-adaptation for finding out the optimum solution to a given problem. These approaches have been popular for quite a long time. Recently, with the availability of massive computational infrastructure, they have emerged as a good option for global search problems. Document clustering is one typical example of the global search problem, where the optimum cluster formation is achieved on the given data.

Of the many available applications of evolutionary clustering algorithms, we will be discussing below the three application areas: image segmentation, text document clustering, and medical data clustering using evolutionary clustering approaches.

2.3.1 Image Segmentation Using Evolutionary Clustering Algorithms

Image segmentation refers to the process of diving an image into multiple parts or segments. It helps in understanding the formation of image in a better way. Various criteria may be used for forming segments, including pixel intensity, color-coding etc. Image segmentation is an important task and has various applications. Some of the prominent applications of image segmentation are machine vision, medical imaging, content-based image retrieval, object detection, object localization, visual surveillance etc. [40]. Various evolutionary approaches have been proposed for image segmentation [41–45]. A quantum evolutionary clustering approach has been proposed by Li et al. [41]. They used this approach to perform image segmentation with the help of watershed algorithm and evolutionary clustering. In another work of Li et al. [42] and by Gong et al. in [43], manifold distance is used with an evolutionary clustering algorithm for image segmentation. Differential evolution is one of the fast, efficient and robust, global search heuristics and has applications in image segmentation. Das et al. [44] proposed an improved differential evolution algorithm.

The particle swarm optimization (PSO)-based image segmentation method proposed in [45] performs better than the popular clustering algorithms like k-means and fuzzy c-means algorithms on different types of image data such as MRI, synthetic images, and satellite images.

2.3.2 Medical and Healthcare Data Clustering Using Evolutionary Clustering Algorithms

The analysis of medical data is becoming more and more important as well as challenging. The newer application areas are demanding that the analysis and presentation of medical and healthcare data should be more intuitive. Medical and healthcare data may include images, structured or unstructured records, text documents etc. Researchers have been trying to devise the innovative ways and means of clustering and segmenting this data. Various evolutionary clustering algorithms exist in literature for medical and healthcare data clustering [46–50]. Li et al. [46] and Maulik et al. [47] have proposed solutions for medical image segmentation using evolutionary approaches. Li et al. [46] used hierarchical clustering and Maulik et al. [47] used a genetic algorithm for the purpose. Marghny et al. [48] used an evolutionary approach for clustering the Hepatitis-C data. The least human intervention and autonomy of operation are the desirable traits of any clustering algorithm. In Šprogar et al. [49], an autonomous evolutionary algorithm has been proposed and used on clustering of medical data. Further, Samanta et al. [50] described some evolutionary algorithms and applications in medical data clustering.

2.3.3 Text Document Clustering Using Evolutionary Clustering Algorithms

Text document clustering is the process of grouping the text documents based on some descriptors. These descriptors may be the words and their frequencies or some other complex presentations. There are many applications of text document clustering, such as World Wide Web (WWW), document retrieval and management, topic extraction etc [51–54]. These applications can be divided into two categories: namely, online and offline document clustering. The online clustering works with the Internet and offline works without the Internet. Different parts of text in a document can also be clustered based on the semantics of the text.

Shi et al. [51] discuss a high-performance genetic algorithm (GA) for text clustering and Karaa et al. [52] also used the variant of GA for document clustering. Some algorithms require the feature selection as an intermediary step before doing the document clustering. The right feature selection helps improve the accuracy of overall task of the clustering. Abualigah et al. [53] proposed a method for feature selection using PSO for document clustering. Song et al [54] discussed a hybrid evolutionary approach for text document clustering. The approach utilizes the benefits of PSO and quantum mechanics for improved performance.

2.4 PERFORMANCE EVALUATION OF EVOLUTIONARY CLUSTERING ALGORITHMS

A plethora of evolutionary approaches exist for clustering. According to the popular "no free lunch" (NFL) theorem – "There exists no single algorithm or a method, which is superior to its competitors in all situations." This means that an algorithm or approach is good for some specific problems and data only. This also applies with regard to the evolutionary clustering approaches. They are good for some specific scenarios only. Genetic algorithm and its variants are good for text document clustering. Quantum-inspired evolutionary clustering approaches are good for medical data clustering. The key advantages of evolutionary algorithms include the flexibility and self-adaptation for finding out the optimum solution to a given problem. These approaches have been popular for quite a long time. Recently, with the availability of massive computational infrastructure, they have emerged as a good option for global search problems.

The clustering techniques can be assessed on external and internal parameters of assessment. For external assessment, we have the prior knowledge about the data and evaluation is done by comparing the clustering performance of actual vs idea cluster formations. Various performance metrics, such as accuracy, f-score, precision, rand index etc., can be used. When the prior knowledge of cluster formation is unavailable then an internal evaluation of clustering is done. There also exist many performance measures and the indices for the internal evaluation of the cluster formation, including the Davies-Bouldin index, the Dunn index, Entropy, Partition Coefficient, the Separation Index, the Silhouette index, Xie and Beni's Index etc.

2.5 CONCLUSION

Evolutionary algorithms are part of a field of study known as nature-inspired computing, which is similar to the biological evolution phenomena of species. There are many types of evolutionary approaches, including Genetic Algorithm, Gene Expression Programming, Differential Evolution, Particle Swarm Optimization, Gray Wolf Optimizer, Antlion Optimizer, Neuroevolutionary Algorithm etc. These approaches are effective in solving the global search optimization problems such as data clustering. This chapter describes the task and categorization of clustering. The key advantages of evolutionary algorithms include the flexibility and self-adaptation for finding out the optimum solution to a given problem. These approaches have been popular for quite some time. Various applications of evolutionary clustering have been discussed, along with the performance evaluation of these algorithms.

REFERENCES

1. Xu, D., & Tian, Y. (2015). A comprehensive survey of clustering algorithms. *Annals of Data Science*, 2(2), 165–193.
2. Rothlauf, F. (2006). Representations for genetic and evolutionary algorithms. In *Representations for Genetic and Evolutionary Algorithms* (pp. 9–32). Springer, Berlin, Heidelberg.
3. Hruschka, E. R., Campello, R. J., & Freitas, A. A. (2009). A survey of evolutionary algorithms for clustering. *IEEE Transactions on Systems, Man, and Cybernetics, Part C (Applications and Reviews)*, 39(2), 133–155.
4. Berkhin, P. (2006). A survey of clustering data mining techniques. In *Grouping Multidimensional Data* (pp. 25–71). Springer, Berlin, Heidelberg.
5. Guyeux, C., Chrétien, S., Bou Tayeh, G., Demerjian, J., & Bahi, J. (2019). Introducing and Comparing Recent Clustering Methods for Massive Data Management in the Internet of Things. *Journal of Sensor and Actuator Networks*, 8(4), 56.
6. Aljarah, I., & Ludwig, S. A. (2012, November). *Parallel particle swarm optimization clustering algorithm based on MapReduce methodology*. In *2012 Fourth World Congress on Nature and Biologically Inspired Computing (NaBIC)* (pp. 104–111). IEEE.
7. Al-Madi, N., Aljarah, I., & Ludwig, S. A. (2014, December). Parallel glowworm swarm optimization clustering algorithm based on MapReduce. In *2014 IEEE Symposium on Swarm Intelligence* (pp. 1–8). IEEE.

8. Aljarah, I., & Ludwig, S. A. (2013, June). *A new clustering approach based on glowworm swarm optimization.* In *2013 IEEE Congress on Evolutionary Computation* (pp. 2642–2649). IEEE.
9. Holland J.H. (1975). *Adaptation in Natural and Artificial System.* The University of Michigan Press, Ann Arbor.
10. Storn, R., & Price, K. (1997). Differential evolution–a simple and efficient heuristic for global optimization over continuous spaces. *Journal of Global Optimization*, 11(4), 341–359.
11. Kennedy, J. and Eberhart, R. (1995). Particle swarm optimization. *Proceedings IEEE International Conference Neural Networks*, 4, 1942–1948.
12. Dorigo, M., & Di Caro, G. (1999). *Ant colony optimization: a new meta-heuristic.* In *Evolutionary Computation, 1999. CEC 99. Proceedings of the 1999 Congress on* (Vol. 2, pp. 1470–1477). IEEE.
13. Karaboga, D. (2005). An idea based on honey bee swarm for numerical optimization (Vol. 200). Technical Report-tr06, Erciyes University, Engineering Faculty, Computer Engineering Department.
14. Mirjalili, S., Mirjalili, S. M., & Lewis, A. (2014). Grey Wolf optimizer. *Advances in Engineering Software*, 69, 46–61.
15. Mirjalili, S., & Lewis, A. (2016). The whale optimization algorithm. *Advances in Engineering Software*, 95, 51–67.
16. Krishnanand, K. N., & Ghose, D. (2006). Glowworm swarm based optimization algorithm for multimodal functions with collective robotics applications. *Multiagent and Grid Systems*, 2(3), 209–222.
17. Rashedi, E., Nezamabadi-Pour, H., & Saryazdi, S. (2009). GSA: A gravitational search algorithm. *Information Sciences*, 179(13), 2232–2248.
18. Yadav, A., Deep, K., Kim, J. H., & Nagar, A. K. (2016). Gravitational swarm optimizer for global optimization. *Swarm and Evolutionary Computation*, 31, 64–89.
19. Formato, R. A. (2007). Central force optimization: a new metaheuristic with applications in applied electromagnetics. *Progress in Electromagnetics Research*, 77, 425–491.
20. Geem, Z.W., Kim, J.H., Loganathan, G.V. (2001). A new heuristic optimization algorithm: harmony search. *Simulation*, 76, 60.
21. Yang, X. S., & Deb, S. (2009, December). *Cuckoo search via Lévy flights.* In *Nature & Biologically Inspired Computing, 2009. NaBIC 2009. World Congress on* (pp. 210–214). IEEE.
22. Yılmaz, S., & Küçüksille, E. U. (2015). A new modification approach on bat algorithm for solving optimization problems. *Applied Soft Computing*, 28, 259–275.
23. Mirjalili, S. (2015). The antlion optimizer. *Advances in Engineering Software*, 83, 80–98.
24. Niknam, T., Amiri, B., Olamaei, J., & Arefi, A. (2009). An efficient hybrid evolutionary optimization algorithm based on PSO and SA for clustering. *Journal of Zhejiang University-SCIENCE A*, 10(4), 512–519.
25. He, Y., Tan, H., Luo, W., Mao, H., Ma, D., Feng, S., & Fan, J. (2011, December). *Mr-DBSCAN: an efficient parallel density-based clustering algorithm using MapReduce.* In *2011 IEEE 17th International Conference on Parallel and Distributed Systems* (pp. 473–480). IEEE.
26. Kisilevich, S., Mansmann, F., & Keim, D. (2010, June). *P-DBSCAN: a density based clustering algorithm for exploration and analysis of attractive areas using collections of geo-tagged photos.* In *Proceedings of the 1st International Conference and Exhibition on Computing for Geospatial Research & Application* (p. 38). ACM.

27. Lv, Y., Ma, T., Tang, M., Cao, J., Tian, Y., Al-Dhelaan, A., & Al-Rodhaan, M. (2016). An efficient and scalable density-based clustering algorithm for datasets with complex structures. *Neurocomputing*, 171, 9–22.
28. Kumar, K. M., & Reddy, A. R. M. (2016). A fast DBSCAN clustering algorithm by accelerating neighbor searching using Groups method. *Pattern Recognition*, 58, 39–48.
29. Wang, L., Zhu, H., Meng, J., & He, W. (2019). Incremental local distribution-based clustering using Bayesian adaptive resonance theory. *IEEE Transactions On Neural Networks and Learning Systems*, 30(11), 3496–3504.
30. Güngör, E., & Özmen, A. (2017). Distance and density based clustering algorithm using Gaussian kernel. *Expert Systems with Applications*, 69, 10–20.
31. Bouguettaya, A., Yu, Q., Liu, X., Zhou, X., & Song, A. (2015). Efficient agglomerative hierarchical clustering. *Expert Systems with Applications*, 42(5), 2785–2797.
32. Reddy, C. K., & Vinzamuri, B. (2014). A survey of partitional and hierarchical clustering algorithms. In *Data Clustering, Ch. 4* (pp. 87–110). Chapman and Hall/CRC, NY.
33. Wang, Y. X., & Xu, H. (2016). Noisy sparse subspace clustering. *The Journal of Machine Learning Research*, 17(1), 320–360.
34. Cao, X., Zhang, C., Fu, H., Liu, S., & Zhang, H. (2015). *Diversity-induced multi-view subspace clustering*. In *Proceedings of the IEEE conference on computer vision and pattern recognition* (pp. 586–594).
35. Dinkar, S.K. & Deep, K. (2019). *Neural Computing & Applications*. doi 10.1007/s00521-019-04174-0.
36. Mukhopadhyay, A., Maulik, U., & Bandyopadhyay, S. (2015). A survey of multiobjective evolutionary clustering. *ACM Computing Surveys (CSUR)*, 47(4), 61.
37. Xu, K. S., Kliger, M., & Hero Iii, A. O. (2014). Adaptive evolutionary clustering. *Data Mining and Knowledge Discovery*, 28(2), 304–336.
38. Dinkar, S. K., & Deep, K. (2017). Opposition based Laplacian ant lion optimizer. *Journal of Computational Science*, 23, 71–90.
39. Dinkar, S. K., & Deep, K. (2018). An efficient opposition based Lévy Flight Antlion optimizer for optimization problems. *Journal of Computational Science*, 29, 119–141.
40. Kushwaha, A. K. S., Sharma, C. M., Khare, M., Prakash, O., & Khare, A. (2014). Adaptive real-time motion segmentation technique based on statistical background model. *The Imaging Science Journal*, 62(5), 285–302.
41. Li, Y., Shi, H., Jiao, L., & Liu, R. (2012). Quantum evolutionary clustering algorithm based on watershed applied to SAR image segmentation. *Neurocomputing*, 87, 90–98.
42. Li, Y., Shi, H., Gong, M., & Shang, R. (2009, June). *Quantum-inspired evolutionary clustering algorithm based on manifold distance*. In *Proceedings of the first ACM/SIGEVO Summit on Genetic and Evolutionary Computation* (pp. 871–874). ACM.
43. Gong, M., Jiao, L., Bo, L., Wang, L., & Zhang, X. (2008). Image texture classification using a manifold-distance-based evolutionary clustering method. *Optical Engineering*, 47(7), 077201.
44. Das, S., Abraham, A., & Konar, A. (2007). Automatic clustering using an improved differential evolution algorithm. *IEEE Transactions on Systems, Man, and Cybernetics-Part A: Systems and Humans*, 38(1), 218–237.
45. Omran, M., Engelbrecht, A. P., & Salman, A. (2005). Particle swarm optimization method for image clustering. *International Journal of Pattern Recognition and Artificial Intelligence*, 19(03), 297–321.
46. Lai, C. C., & Chang, C. Y. (2009). A hierarchical evolutionary algorithm for automatic medical image segmentation. *Expert Systems with Applications*, 36(1), 248–259.
47. Maulik, U. (2009). Medical image segmentation using genetic algorithms. *IEEE Transactions on Information Technology in Biomedicine*, 13(2), 166–173.

48. Marghny, M. H., El-Aziz, R. M. A., & Taloba, A. I. (2014). An effective evolutionary clustering algorithm: Hepatitis C case study. arXiv preprint arXiv:1405.6173.
49. Šprogar, M., Šprogar, M., & Colnarič, M. (2005). Autonomous evolutionary algorithm in medical data analysis. *Computer Methods and Programs in Biomedicine*, 80, S29–S38.
50. Samanta, Sourav, Choudhury, Alkoparna, Dey, N., Ashour, A. S., & Balas, V. E. (2017). Quantum-inspired evolutionary algorithm for scaling factor optimization during manifold medical information embedding. In *Quantum Inspired Computational Intelligence, Ch. 9* (pp. 285–326). Morgan Kaufmann.
51. Shi, K., & Li, L. (2013). High performance genetic algorithm based text clustering using parts of speech and outlier elimination. *Applied Intelligence*, 38(4), 511–519.
52. Karaa, W. B. A., Ashour, A. S., Sassi, D. B., Roy, P., Kausar, N., & Dey, N. (2016). Medline text mining: an enhancement genetic algorithm based approach for document clustering. In *Applications of Intelligent Optimization in Biology and Medicine* (pp. 267–287). Springer, Cham.
53. Abualigah, L. M., Khader, A. T., & Hanandeh, E. S. (2018). A new feature selection method to improve the document clustering using particle swarm optimization algorithm. *Journal of Computational Science*, 25, 456–466.
54. Song, W., Qiao, Y., Park, S. C., & Qian, X. (2015). A hybrid evolutionary computation approach with its application for optimizing text document clustering. *Expert Systems with Applications*, 42(5), 2517–2524.

3 Solving Linear Fractional Programming Problem Using Revised and Column Simplex Method

Ather Aziz Raina, Srikant Gupta, Irfan Ali, and Javid Iqbal

CONTENTS

3.1 INTRODUCTION

In Linear Fractional Programming (LFP), the objective function is a ratio of two linear functions instead of a single linear function (as is the case with Linear Programming). Thus, instead of computing the best outcome such as maximum profit or lowest cost, the highest ratio of outcome to cost is computed. This ratio represents the highest efficiency. Fractional Linear Programming (FLP) problem is a special kind of non-linear programming problem which is quite commonly used in finance, industries, health care and production planning. Both linear programming and linear-fractional programming represent optimization problems using linear equations and linear inequalities, which, for each problem-instance, define a feasible set. Fractional linear programs have a more well-appointed set of objective functions.

DOI: 10.1201/9781003089636-3

In the present investigation, we consider two approaches for solving a Linear Fractional Programming problem. These approaches are based on a revised and column simplex method and incorporate many remarkable features which were not taken together by other researchers working on the same lines. The critical concepts incorporated for solving the linear fractional programming problem is (i) revised simplex method (ii) column simplex method. The remaining content of this chapter is structured in different sections. In Section 3.2, we have provided a review of the related works done by the eminent researchers. In Section 3.3, the mathematical formulation of LP and LFP are presented. The solution of linear fractional programming problem using revised simplex method is given in Section 3.4. In Section 3.5, solution of linear fractional programming problem using column simplex method is presented. In the final Sections 3.6 and 3.7, comparison and conclusion has been drawn which summarizes the research investigation.

3.2 LITERATURE REVIEW

The Hungarian mathematician (Martos, 1960; Martos and Whinston, 1964) was the first to develop the field of linear fractional programming problem. Various methods are proposed to solve this problem. Tantawy (2008a) proposed a suitable methodology and the primary concept behind this technique to solve the LFP problem by searching the feasible region through a series of points in the direction that optimizes the value of the objective function. Tantawy (2008b) creates another LFP solution method that can be used to sensitivity analysis. Chergui and Moulaï (2008) to optimize fractional objective functions over a subset of the original continuous feasible set then, if necessary, a branching process is performed until a feasible integer solution is obtained. Pramanik et al. (2011) proposed a technique for solving multi-objective linear plus linear fractional programming problem based on approximation of the Taylor series. Tantawy (2014) also suggested an approach to duality to solve the problem of linear fractional programming. Ait Mehdi et al. (2014) described an improvement in the method used by a new homotopy perturbation method developed by Das and Mandal (2015) to find the exact solutions with equality restrictions for the system of LFP problem. Saha et al. (2015) researched an approach to solving LFP by transforming it into a single LP problem that can be solved using any method and developed FORTRAN programs to solve the problem. Simi and Talukder (2017) provided a unique approach to solving the problem of LFP by transforming the problem into a linear programming (LP) problem, and they solved the converted problem using the concept of duality. Gupta et al. (2017) suggested an iterative procedure based on Beal's technique to solve the LFP problem. The suggested technique has the benefit of avoiding the complexity of conventional procedures; it can be easily used in a minimum number of iterations to solve any FLP problem. Sivri et al. (2018) proposed a solution to the quadratic FP problems involving an objective factorized or non-factorized function and subject to homogeneous or non-homogenous constraints. The suggested method produces consecutive LP problems and specifies an optimal solution to the problem of quadratic FP. Jain et al. (2018) provided a complete ranking and scanning of the integer feasible solutions of quadratic fractional integer programming problems with restricted variables by establishing the existence of a linear or linear fractional function that acts as a lower limit on the objective function values over the entire feasible set. By using the properties of dualizing parametrization functions, Lagrangian functions and the epigraph of conjugate functions, Sun et al. (2018) presented Farkas-type results for a fractional programming problem. In considering the problem of optimizing the ratio of two convex functions over a closed and convex set in the space of matrices that can be categorized as a double-convex fractional programming problem, Bouhamidi et al. (2018) suggested a conditional gradient approach.

3.3 METHODOLOGY

The following mathematical methodology has been used to solve the LFP problem.

3.3.1 Definition and Mathematical Formulation of LP and LFP

The mathematical expression of a general linear programming problem (in matrix notation) is as follows:

Find $x_1, x_2, x_3, \ldots, x_n$, so as to optimize

$$\text{Maximize}\left(\text{or Minimize}\right) Z = CX$$

$$\text{Subject to} \qquad AX\left(\le, =, \ge\right)b$$

$$X \ge 0; b \ge 0,$$

where $A = [a_{ij}]_{m \times n}$ called the coefficient matrix, C is an $1 \times n$ row matrix, **b** is an $m \times 1$ column matrix, X is an $n \times 1$ column matrix and **0** is a null matrix of the type $n \times 1$.

A linear fractional programming problem occurs when a linear fractional function is to be maximized and the problem can be formulated mathematically (in matrix notation) is as follows:

$$\text{Maximize}\left(\text{or Minimize}\right) Q\left(X\right) = \frac{p\left(X\right)}{d\left(X\right)} = \frac{C^T X + \alpha}{D^T X + \beta}$$

$$\text{Subject to} \qquad A\underline{X}\left(\le, =, \ge\right)b,$$

$$\underline{X} \ge 0;\ b \ge 0,$$

where $A = [a_{ij}]_{m \times n}$ called the coefficient matrix, C is an $1 \times n$ row matrix, **b** is an $m \times 1$ column matrix, X is an $n \times 1$ column matrix, $\mathbf{b} \in R^m$, $X, C, D \in R^n$, $\alpha, \beta \in R$.

3.3.2 Revised Simplex Method for Solving LFP Problem

The revised simplex method is an improvisation of the simplex method. In the simplex method, all the entries are calculated to complete a simplex table, whereas in the revised simplex method, only the necessary information required for computing the current simplex iteration is updated and stored. The revised simplex method is a powerful tool to solve many problems that involve a large number of decision variables and constraints, such as inventory, diet, product mix, and trim loss problems.

The following steps are used for solving the fractional linear programming problem by revised simplex method:

Step 1: Find an initial basic feasible solution with B as basis matrix. Compute B^{-1}.

Step 2: Calculate $w_p = C'_{BP}B^{-1}$, $w_d = C'_{Bd}B^{-1}$ and $\underline{\bar{b}} = B^{-1}\underline{b}$.

Step 3: Set up the following revised simplex tableau

	RHS	
x_{B1}		
$\vdots$	$\underline{\bar{b}}$	B^{-1}
x_{Bn}		
$p(X)$	α	$\underline{w}_p$
$d(X)$	β	$\underline{w}_d$
$Q(X)=\dfrac{p(X)}{d(X)}$		

where $\underline{\bar{b}}$ is the cell representing the current value of the basic variables.

Step 4: For each non-basic variables calculate $Z_j - C_j$.

Note that with $\underline{w}_p = \underline{C}'_{BP}B^{-1}$ & $\underline{w}_d = \underline{C}'_{Bd}B^{-1}$, $z_j = \underline{C}'_{BP}B^{-1}a_j = w_P a_j$ and $\left(Z_j - C_j\right)_p = \underline{w}_P a_j - C_j$ similarly $\left(Z_j - C_j\right)_d = \underline{w}_d a_j - C_j$. As the objective function is in the form of fraction we have to calculate $\Delta_i = (Z_j - C_j)_p - Q(Z_j - C_j)_d$.

Step 5: Let $\Delta_k = \min(\Delta_i)$. If $\Delta_k \geq 0$ (for maximization of FLPP) STOP; the current basic feasible solution is optimal, otherwise go to step 6.

Step 6: Calculate $y_k = B^{-1}a_k$. If $y_k \leq 0$, STOP; the given FLPP has an unbounded solution, otherwise go to step 7.

$$y_k$$
$$(Z_j - C_j)_p$$
$$(Z_j - C_j)_d$$

Step 7: Insert the column to the right of the tableau given at step 4.

Leading to the following tableau

		RHS	
x_{B1}	$\underline{b}$		y_k
$\vdots$			
x_{Bn}	B^{-1}		
$p(X)$	α	$\underline{w}_p$	$(Z_j - C_j)_p$
$d(X)$	β	$\underline{w}_d$	$(Z_j - C_j)_d$
$Q(X) = \dfrac{p(X)}{d(X)}$			

Step 8: Determine the ratio as follows $\dfrac{\bar{b}_r}{y_{rk}}\left\{\min_i \dfrac{\bar{b}_r}{y_{rk}} | y_{ik} \geq 0\right\}$.

Step 9: Update the revised simplex tableau using pivoting operations. After pivoting remove the additional column corresponding to x_k and x_k enter in place x_{Br} of as basic variables. This will give the updated tableau for the next iteration. Repeat step 4 onwards with updated tableau.

3.3.3 Column Simplex Method for Solving LFP Problem

The technique of column simplex is regarded to be the expansion of the technique of classical simplex. The main advantage connected with the column simplex method is that all calculations are based on a column-wise iteration, while all calculations rely on row-wise iterations in the classical simplex method. No author used this technique to solve their issue with the best of my understanding. The following procedure is required for solving the fractional linear programming problem by column simplex method:

To apply the column simplex method let us consider the FLP problem as

$$Maximize(\text{or } Minimize)Q(X) = \frac{p(X)}{d(X)} = \frac{C^T X + \alpha}{D^T X + \beta}$$

$$\text{Subject to} \qquad A\underline{X}(\leq, =, \geq)b,$$

$$\underline{X} \geq 0; b \geq 0,$$

where $A = [a_{ij}]_{m \times n}$ called the coefficient matrix, C is an $1 \times n$ row matrix, **b** is an $m \times 1$ column matrix, X is an $n \times 1$ column matrix, $\mathbf{b} \in R^m$, $X, C, D \in R^n$, $\alpha, \beta \in R$. The starting column simplex tableau for FLPP is usually written as:

		$\underline{x}_j$	
	Q	α/β	Δ_j
p		α	$-\underline{c}_p$
d		β	$-\underline{c}_d$
$\underline{X}$		β	$-I$
$\underline{X}_S$		$\underline{b}$	A

The starting solution is $\underline{X} = 0$ and $\underline{X}_S = \underline{b}$ with $Q = \frac{\alpha}{\beta}$, $\underline{X}_S$ denote the slack variables. As the objective function is in the form of fraction so we have calculate for each of the non-basic variables $\Delta_i = (Z_j - C_j)_p - Q(Z_j - C_j)_d$. If $\Delta_k \geq 0$ (for maximization of FLPP) STOP; the current basic feasible solution is optimal while for minimization problem Δ_i should be ≤ 0.

3.4 NUMERICAL ILLUSTRATION

3.4.1 Numerical Examples Based on Revised Simplex Method

In this section, we numerically illustrate two examples to demonstrate our proposed method of revised simplex method.

Example 1

Consider the Fractional Linear Programming (Flp) Problem

$$MaxQ(X) = \frac{p(X)}{d(X)} = \frac{6x_1 + 3x_2 + 6}{5x_1 + 2x_2 + 5}$$

$$\text{s.t} \quad 4x_1 + 2x_2 \leq 20$$
$$3x_1 + 5x_2 \leq 25$$
$$x_1, x_2 \geq 0$$

The given problem reduces to the following form:

$$MaxQ(X) = \frac{p(X)}{d(X)} = \frac{6x_1 + 3x_2 + 0x_3 + 0x_4 + 6}{5x_1 + 2x_2 + 0x_3 + 0x_4 + 5}$$

$$\text{s.t} \quad 4x_1 + 2x_2 + x_3 + 0x_4 = 20$$
$$3x_1 + 5x_2 + 0x_3 + x_4 = 25$$
$$x_1, x_2, x_3, x_4 \geq 0$$

The starting basic feasible solution is $\underline{X}_B = \begin{pmatrix} x_3 \\ x_4 \end{pmatrix} = \underline{b} = \begin{pmatrix} 20 \\ 25 \end{pmatrix}$ with $Q(x) = \frac{6}{5}$ and the basis matrix $B = \begin{pmatrix} 1 & 0 \\ 0 & 1 \end{pmatrix} = B^{-1}$.

The starting revised simplex tableau is

Tableau(1)

x_3	20	1	0
x_4	25	0	1
p	6	0	0
d	5	0	0
$Q = 6/5$			

Note that $\underline{C}'_{BP} = (0,0)$, $\underline{C}'_{Bd} = (0,0)$,$\bar{b} = \begin{pmatrix} 20 \\ 25 \end{pmatrix}$,$\underline{C}'_{Bp}\bar{b} = 0$,$\underline{C}'_{Bd}\bar{b} = 0$, $\underline{w}'_p = \underline{C}'_{BP}B^{-1} = \underline{0}$, $\underline{w}'_d = \underline{C}'_{Bd}B^{-1} = \underline{0}$. x_1 and x_2 are the non-basic variables. Thus,

$$(Z_1 - C_1)_p = \underline{w}'_p a_1 - C_1 = (00)\begin{pmatrix}4\\3\end{pmatrix} - 6 = -6$$

$$(Z_1 - C_1)_d = \underline{w}'_d a_1 - C_1 = (00)\begin{pmatrix}4\\3\end{pmatrix} - 5 = -5$$

$$\Rightarrow \Delta_1 = -6 + \frac{6}{5} \times 5 = 0$$

$$(Z_2 - C_2)_p = \underline{w}'_p a_2 - C_2 = (00)\begin{pmatrix}-2\\5\end{pmatrix} - 3 = -3$$

$$(Z_2 - C_2)_d = \underline{w}'_d a_2 - C_2 = (00)\begin{pmatrix}-2\\5\end{pmatrix} - 2 = -2$$

$$\Rightarrow \Delta_2 = -3 + \frac{6}{5} \times 2 = \frac{-3}{5}$$

$$\text{Now } \Delta_k = \min(\Delta_1, \Delta_2) = \frac{-3}{5} < 0$$

$\Rightarrow$ x_2 will become basic and $k = 2$.

Now $\underline{y}_k = \underline{y}_2 = B^{-1}a_2 = \begin{pmatrix}1 & 0\\0 & 1\end{pmatrix}\begin{pmatrix}-2\\5\end{pmatrix} = \begin{pmatrix}-2\\5\end{pmatrix}$ This gives the additional column corresponding to x_2. The tableau becomes:

Tableau (1′) x_2

x_3	20	1	0	−2	-
x_4	25	0	1	5	25/5=5
					<-
p	6	0	0	−3	
d	5	0	0	−2	

Pivoting at $y_{22} = 5$ gives the updated tableau as:

Tableau (2)

x_3	30	1	2/5
x_2	5	0	1/5
p	21	0	3/5
d	15	0	2/5
$Q = 21/15$			

The current non-basic variables are x_1 and x_4. Thus

$$(Z_1 - C_1)_p = \underline{w}'_p a_1 - C_1 = \left(0\,\frac{3}{5}\right)\begin{pmatrix}4\\3\end{pmatrix} - 6 = -\frac{21}{5}$$

$$(Z_1 - C_1)_d = \underline{w}_d' a_1 - C_1 = \begin{pmatrix} 0 & \frac{2}{5} \end{pmatrix} \begin{pmatrix} 4 \\ 3 \end{pmatrix} - 5 = -\frac{19}{5}$$

$$\Rightarrow \Delta_1 = -\frac{21}{5} + \frac{21}{5} \times \frac{19}{5} = \frac{84}{75} > 0$$

$$(Z_4 - C_4)_p = \underline{w}_p' a_4 - C_4 = \begin{pmatrix} 0 & \frac{3}{5} \end{pmatrix} \begin{pmatrix} 0 \\ 1 \end{pmatrix} - 0 = -\frac{3}{5}$$

$$(Z_4 - C_4)_d = \underline{w}_d' a_4 - C_4 = \begin{pmatrix} 0 & \frac{2}{5} \end{pmatrix} \begin{pmatrix} 0 \\ 1 \end{pmatrix} - 0 = -\frac{2}{5}$$

$$\Rightarrow \Delta_4 = \frac{3}{5} - \frac{21}{5} \times \frac{2}{5} = \frac{3}{75} > 0$$

The current solution is optimum and the procedure terminates. The optimal solution given by the tableau is $x_1 = 0$, $x_2 = 5$ and $Q = \frac{21}{15}$.

Example 2

Consider the fractional linear programming (FLP) problem

$$Max Q(X) = \frac{p(X)}{d(X)} = \frac{8x_1 + 9x_2 + 4x_3 + 4}{2x_1 + 3x_2 + 2x_3 + 7}$$

$$\begin{aligned} \text{s.t} \quad & x_1 + x_2 + 2x_3 \le 3 \\ & 2x_1 + x_2 + 4x_3 \le 4 \\ & 5x_1 + 3x_2 + x_3 \le 15 \\ & x_1, x_2, x_3 \ge 0 \end{aligned}$$

The given problem reduces to the following form:

$$Max Q(X) = \frac{p(X)}{d(X)} = \frac{8x_1 + 9x_2 + 4x_3 + 0x_4 + 0x_5 + 4}{2x_1 + 3x_2 + 2x_3 + 0x_4 + 0x_5 + 7}$$

$$\begin{aligned} \text{s.t} \quad & x_1 + x_2 + 2x_3 + 0x_4 + 0x_5 + 0x_6 = 3 \\ & 2x_1 + x_2 + 4x_3 + 0x_4 + 0x_5 + 0x_6 = 4 \\ & 5x_1 + 3x_2 + x_3 + 0x_4 + 0x_5 + 0x_6 = 15 \\ & x_1, x_2, x_3, x_4, x_5, x_6 \ge 0 \end{aligned}$$

The starting basic feasible solution is $\underline{X}_B = \begin{pmatrix} x_3 \\ x_4 \\ x_5 \end{pmatrix} = \underline{b} = \begin{pmatrix} 3 \\ 4 \\ 15 \end{pmatrix}$ with $Q = 4/7$ and the basis matrix $B = \begin{pmatrix} 1 & 0 & 0 \\ 0 & 1 & 0 \\ 0 & 0 & 1 \end{pmatrix} = B^{-1}$ also.

The starting revised simplex tableau is

Tableau (1)

x_4	3	1	0	0
x_5	4	0	1	0
x_6	15	0	0	1
p	4	0	0	0
d	7	0	0	0
$Q = 4/7$				

Note that: $\underline{C}'_{Bp} = (0,0,0)$, $\underline{C}'_{Bd} = (0,0,0)$, $\bar{b} = \begin{pmatrix} 3 \\ 4 \\ 15 \end{pmatrix}$, $\underline{C}'_{Bp}\bar{b} = 0$, $\underline{C}'_{Bd}\bar{b} = 0$,

$\underline{w}'_p = \underline{C}'_{Bp}B^{-1} = \underline{0}$, $\underline{w}'_d = \underline{C}'_{Bd}B^{-1} = \underline{0}$. x_1, x_2, x_3 are the non-basic variables. Thus,

$$(Z_1 - C_1)_p = \underline{w}'_p a_1 - C_1 = (000)\begin{pmatrix} 1 \\ 2 \\ 5 \end{pmatrix} - 8 = -8$$

$$(Z_1 - C_1)_d = \underline{w}'_d a_1 - C_1 = (000)\begin{pmatrix} 1 \\ 2 \\ 5 \end{pmatrix} - 2 = -2$$

$$\Delta_1 = -8 + \frac{4}{7} \times 2 = \frac{-50}{7}$$

$$(Z_2 - C_2)_p = \underline{w}'_p a_2 - C_2 = (000)\begin{pmatrix} 1 \\ 1 \\ 3 \end{pmatrix} - 9 = -9$$

$$(Z_2 - C_2)_d = \underline{w}'_d a_2 - C_2 = (000)\begin{pmatrix} 1 \\ 1 \\ 3 \end{pmatrix} - 3 = -3$$

$$\Delta_2 = -9 + \frac{4}{7} \times 3 = \frac{-51}{7}$$

$$(Z_3 - C_3)_p = \underline{w}'_p a_3 - C_3 = (000)\begin{pmatrix} 2 \\ 4 \\ 1 \end{pmatrix} - 4 = -4$$

$$(Z_3 - C_3)_d = \underline{w}'_d a_3 - C_3 = (000)\begin{pmatrix} 2 \\ 4 \\ 1 \end{pmatrix} - 2 = -2$$

$$\Delta_3 = -4 + \frac{4}{7} \times 2 = \frac{-20}{7}$$

$$\text{Now } \Delta_k = \min(\Delta_1, \Delta_2, \Delta_3) = \frac{-51}{7}$$

$\Rightarrow$ x_2 will become basic and $k = 2$

Now $\underline{y}_k = \underline{y}_2 = B^{-1}a_2 = \begin{pmatrix} 1 & 0 & 0 \\ 0 & 1 & 0 \\ 0 & 0 & 1 \end{pmatrix}\begin{pmatrix} 1 \\ 1 \\ 3 \end{pmatrix} = \begin{pmatrix} 1 \\ 1 \\ 3 \end{pmatrix}$ this gives the additional column corresponding to x_2.

The tableau becomes:

Tableau (1′)

					x_2	
x_4	3	1	0	0	1	3/1 = 3←
x_5	4	0	1	0	1	4/1 = 4
x_6	15	0	0	1	3	15/3 = 5
p	4	0	0	0	−9	
d	7	0	0	0	−3	

Pivoting at $y_{11} = 1$ gives the updated tableau as:

Tableau (2)

x_2	3	1	0	0
x_5	1	−1	1	0
x_6	6	−3	0	1
p	31	9	0	0
d	16	3	0	0
$Q = 31/16$				

The current non basic variables are x_1, x_3, x_4. Thus

$$(Z_1 - C_1)_p = \underline{w}_p' a_1 - C_1 = (900)\begin{pmatrix} 1 \\ 2 \\ 5 \end{pmatrix} - 8 = 1$$

$$(Z_1 - C_1)_d = \underline{w}_d' a_1 - C_1 = (300)\begin{pmatrix} 1 \\ 2 \\ 5 \end{pmatrix} - 2 = 1$$

$$\Delta_1 = 1 - \frac{31}{16} \times 1 = \frac{-15}{16}$$

$$(Z_3 - C_3)_p = \underline{w}_p' a_3 - C_3 = (900)\begin{pmatrix} 2 \\ 4 \\ 1 \end{pmatrix} - 4 = 14$$

$$(Z_3 - C_3)_d = \underline{w}_d' a_3 - C_3 = (300)\begin{pmatrix} 2 \\ 4 \\ 1 \end{pmatrix} - 2 = 4$$

$$\Delta_3 = 14 - \frac{31}{16} \times 4 = \frac{25}{4}$$

$$(Z_4 - C_4)_p = \underline{w}_p' a_4 - C_4 = (900)\begin{pmatrix} 1 \\ 0 \\ 0 \end{pmatrix} - 0 = 9$$

$$(Z_4 - C_4)_d = \underline{w}_d' a_4 - C_4 = (300)\begin{pmatrix} 1 \\ 0 \\ 0 \end{pmatrix} - 0 = 3$$

$$\Delta_4 = 9 - \frac{31}{16} \times 3 = \frac{51}{16}$$

Now $\Delta_k = \min(\Delta_1, \Delta_3, \Delta_4) = -\frac{15}{16}$

$\Rightarrow x_1$ will become basic and $k = 1$

Now $\underline{y}_k = \underline{y}_1 = B^{-1}a_1 = \begin{pmatrix} 1 & 0 & 0 \\ -1 & 1 & 0 \\ -3 & 0 & 1 \end{pmatrix}\begin{pmatrix} 1 \\ 2 \\ 5 \end{pmatrix} = \begin{pmatrix} 1 \\ 1 \\ 2 \end{pmatrix}$. This gives the additional column corresponding to x_1. The tableau becomes

Tableau (2′)

				x_1		
x_2	3	1	0	0	1	3/1 = 3
x_5	1	−1	1	0	1	1/1 = 1←
x_6	6	−3	0	1	2	6/2 = 3
p	31	9	0	0	1	
d	16	3	0	0	1	

Pivoting at $y_{22} = 1$ gives the updated tableau as:

Tableau (3)

x_2	2	2	−1	0
x_1	1	−1	1	0
x_6	4	−1	−2	1
p	30	10	−1	0
d	15	4	−1	0
$Q = 30/15$				

The current non-basic variables are x_3, x_4, x_5. Thus,

$$(Z_3 - C_3)_p = \underline{w}_p' a_3 - C_3 = (10 - 10)\begin{pmatrix} 2 \\ 4 \\ 1 \end{pmatrix} - 4 = 12$$

$$(Z_3 - C_3)_d = \underline{w}_d' a_3 - C_3 = (4 - 10)\begin{pmatrix} 2 \\ 4 \\ 1 \end{pmatrix} - 2 = 2$$

$$\Delta_3 = 12 - \frac{30}{15} \times 2 = 8 > 0$$

$$(Z_4 - C_4)_p = \underline{w}_p' a_4 - C_4 = (10 - 10)\begin{pmatrix} 1 \\ 0 \\ 0 \end{pmatrix} - 0 = 10$$

$$(Z_4 - C_4)_d = \underline{w}_d' a_4 - C_4 = (4 - 10)\begin{pmatrix} 1 \\ 0 \\ 0 \end{pmatrix} - 0 = 4$$

$$\Delta_4 = 10 - \frac{30}{15} \times 4 = 2 > 0$$

$$\left(Z_5 - C_5\right)_p = \underline{w}_p' a_5 - C_5 = \left(10 - 10\right)\begin{pmatrix} 0 \\ 1 \\ 0 \end{pmatrix} - 0 = -1$$

$$\left(Z_5 - C_5\right)_d = \underline{w}_d' a_5 - C_5 = \left(4 - 10\right)\begin{pmatrix} 0 \\ 1 \\ 0 \end{pmatrix} - 0 = -1$$

$$\Delta_5 = -1 + \frac{30}{15} \times 1 = 1 > 0$$

The current solution is optimum and the procedure terminates. The optimal solution given by the tableau is $x_1 = 1$, $x_2 = 2$, $x_3 = 0$ and $Q = 30/15$.

3.4.2 Numerical Examples of Column Simplex Method

In this section, we numerically illustrate two examples to demonstrate our proposed method of the column simplex method.

Example 1

Consider the fractional linear programming (FLP) problem

$$MaxQ(X) = \frac{p(X)}{d(X)} = \frac{6x_1 + 3x_2 + 6}{5x_1 + 2x_2 + 5}$$

$$\begin{aligned} \text{s.t}\quad & 4x_1 + 2x_2 \le 20 \\ & 3x_1 + 5x_2 \le 25 \\ & x_1, x_2 \ge 0 \end{aligned}$$

We have

$$\underline{C}_p = (63),\ \underline{C}_d = (52),\ A = \begin{pmatrix} 4 & -2 \\ 3 & 5 \end{pmatrix},\ \underline{X} = (x_1 x_2),\ \underline{X_S} = (x_3 x_4)$$

The starting column simplex tableau is given as

Tableau (1)

		x_1	x_2	
Q	6/5	0	−3/5	
p	6	−6	−3	
d	5	−5	−2	
x_1	0	−1	0	–
x_2	0	0	−1	–
x_3	20	4	−2	–
x_4	25	3	5	25/5 = 5←
			↑	

$\Delta_1 = -6 + \frac{6}{5}5 = 0$, $\Delta_2 = -3 + \frac{6}{5}2 = \frac{-3}{5}$. The most negative $\Delta_i = \Delta_2 = \frac{-3}{5}$ which correspond to x_2. $\Rightarrow x_2$ will become basic and the column corresponding to x_2 will be the pivotal column. The minimum positive ratio of the present solution column to the pivotal column is attained corresponding to x_4.

⇒ the x_4 row will be the pivotal row.

x_4 will become non-basic and the element 5 at the intersection of the pivotal row and the pivotal column will be the pivotal element. The next iteration is obtained by

(1) Dividing the pivotal column by –5.
(2) Applying the elimination procedure to transform the pivotal row (25, 3, 5) into (0, 0, –1).

The subsequent tableau of the column simplex method is:

Tableau (2)

		x_1	x_2
Q	21/15	84/75	3/75
p	21	–21/5	3/5
d	15	–19/5	2/5
x_1	0	–1	0
x_2	5	3/5	1/5
x_3	30	26/5	2/5
x_4	0	0	–1

$\Delta_1 = \dfrac{-21}{5} + \dfrac{21}{15} \times \dfrac{19}{5} = \dfrac{84}{75} > 0$, $\Delta_4 = \dfrac{3}{5} - \dfrac{21}{15} \times \dfrac{2}{5} = \dfrac{3}{75} > 0$. Since all $\Delta_i \geq 0$, the current solution is the required optimal solution. $x_1 = 0$, $x_2 = 5$, $Q = \dfrac{21}{15}$.

Example 2

Consider the fractional linear programming (FLP) problem

$$MaxQ(X) = \frac{p(X)}{d(X)} = \frac{2x_1 + x_2}{3x_1 + x_2 + 6}$$

$$\begin{aligned} \text{s.t} \quad & 5x_1 + 3x_2 \leq 6 \\ & 7x_1 + x_2 \leq 6 \\ & x_1, x_2 \geq 0 \end{aligned}$$

We have

$$\underline{C}_p = (21),\ \underline{C}_d = (71),\ A = \begin{pmatrix} 5 & 3 \\ 7 & 1 \end{pmatrix},\ \underline{X} = (x_1 x_2),\ \underline{X_S} = (x_3 x_4)$$

The starting column simplex tableau is given as

Tableau (1)

		x_1	x_2	
Q	0	–2	–1	
p	0	–2	–1	
d	6	–3	–1	
x_1	0	–1	0	–
x_2	0	0	–1	–
x_3	6	5	3	6/5
x_4	6	7	1	6/7←
			↑	

$\Delta_1 = -2 + 0 \times 3 = -2$, $\Delta_2 = -1 + 0 \times 1 = -1$. The most negative $\Delta_i = \Delta_1 = -2$ which correspond to x_1. $\Rightarrow x_1$ will become basic and the column corresponding to x_1 will be the pivotal column. The minimum positive ratio of the present solution column to the pivotal column is attained corresponding to x_4.

$\Rightarrow$ the x_4 row will be the pivotal row. x_4 will become non-basic and the element 7 at the intersection of the pivotal row and the pivotal column will be the pivotal element.

The next iteration is obtained by

(1) Dividing the pivotal column by –7.
(2) Applying the elimination procedure to transform the pivotal row $(6,7,1)$ into $(0,-1,0)$.

The subsequent tableau of the column simplex method is:

Tableau (2)

		x_4	x_2	
Q	1/5	1/5	−3/5	
p	12/7	2/7	−5/7	
d	60/7	3/7	−4/7	
x_1	6/7	1/7	1/7	6/7
x_2	0	0	−1	–
x_3	12/7	−5/7	16/7	12/16←
x_4	0	−1	0	–
			↑	

$\Delta_4 = \frac{2}{7} - \frac{1}{5} \times \frac{3}{7} = \frac{1}{5}$, $\Delta_2 = \frac{-5}{7} + \frac{1}{5} \times \frac{4}{7} = \frac{-3}{5}$. The most negative $\Delta_i = \Delta_2 = -\frac{3}{5}$ which correspond to x_2. $\Rightarrow x_2$ will become basic and the column corresponding to x_2 will be the pivotal column. The minimum positive ratio of the present solution column to the pivotal column is attained corresponding to x_3.

$\Rightarrow$ the x_3 row will be the pivotal row.

x_3 will become non basic and the element 16/7 at the intersection of the pivotal row and the pivotal column will be the pivotal element.

The next iteration is obtained by

(1) Dividing the pivotal column by −16/7
(2) Applying the elimination procedure to transform the pivotal row $\left(\frac{12}{7}, \frac{-5}{7}, \frac{16}{7}\right)$ into $(0,0,-1)$.

The subsequent tableau of the column simplex method is:

Tableau (3)

		x_4	x_3
Q	1/4	0	1/4
p	9/4	1/16	5/16
d	9	1/4	1/4
x_1	3/4	3/16	−1/16
x_2	3/4	−5/16	7/16
x_3	0	0	−1
x_4	0	−1	0

$\Delta_4 = \frac{1}{16} - \frac{1}{4} \times \frac{1}{4} = 0$, $\Delta_3 = \frac{5}{16} - \frac{1}{4} \times \frac{1}{4} = \frac{1}{4}$. Since all $\Delta_i \geq 0$, the optimal continuous solution is given by $x_1 = \frac{3}{4}$, $x_2 = \frac{3}{4}$, $Q = \frac{1}{4}$.

3.5 COMPARISON AND DISCUSSION

In this section, we compare our results with the existing approach and we find that our results are better than the existing approaches. The reasons are as follows:

- Any type of LFP problem can be solved by this method.
- The LFP problem can be solved easily without transforming it into the LP problem.
- In some cases of complicated numerator and denominator, other existing methods are failed but our method can solve any kind of problem easily.
- Compared to techniques used to solve LFP, its computational steps are simple.
- The final result converges quickly in this method.

Table 3.1 shows the results of conversion needed before solving the LFP problem into LP problem or not.

In Table 3.2, we have compared our results with the previous results and found that in all the cases our number of iteration used in solving the numerical is less than

TABLE 3.1
Comparison of Methods

Method	Charnes and Cooper (1962)	Hasan and Acharjee (2011)	Pandian and Jayalakshmi (2013)	Shah et al. (2015)	Simi and Talukder (2017)	Proposed
Example 1	Required	Required	Required	Required	Required	Not Required
Example 2	Required	Required	Required	Required	Required	Not Required

TABLE 3.2
Comparison of Results

Method	Methods	Objective Function Value	Decision Variable	Iteration	Time
Revised Simplex Method	Shen et al. (2017)	1.6232	(0.0,0.2817)	43	4.322s
	Zhang and Wang (2017)	1.6232	(0.0,0.2839)	65	0.953s
	Liu et al. (2019)	1.6232	(0.0,0.2839)	1983	40.353s
	Proposed Approach	1.6232	(0.0,0.2817)	2	720.012s
Column Simplex Method	Shen and Wang (2006)	3.2916	(3.0, 4.0)	9	0.00004s
	Jiao et al. (2016)	3.2916	(3.0, 4.0)	2	0.0017s
	Liu et al. (2019)	3.2916	(3.0, 4.0)	78	0.00006s
	Liu et al. (2019)	3.2916	(3.0, 4.0)	693	16.5359s
	Proposed Approach	3.2916	(3.0, 4.0)	3	829.97s

as compared to other methods. One of the main advantages of our proposed method is that we can solved it directly without converting it into linear programming problem.

3.6 CONCLUSION

In this chapter, we have presented an improved method for solving LFP problem our approach generates the whole efficient solution for LFP problems. The proposed method is not based upon classical simplex method for solving the LFP problem which searches along the boundary from one feasible vertex to an adjacent vertex until the optimal solution is found and this is the thing which makes it different from all earlier methods of solving LFP problem. The main advantage of the proposed method is that it does not need the given problem to be converted into an LP problem. This proposed method can easily solve any type of large scale LFP problem with ease. Thus our proposed method saves time and energy as it is easy to implement.

REFERENCES

Ait Mehdi, M., Chergui, M. E. A., and Abbas, M. (2014). An improved method for solving multiobjective integer linear fractional programming problem. *Advances in Decision Sciences*, 2014. http://dx.doi.org/10.1155/2014/306456.

Bouhamidi, A., Bellalij, M., Enkhbat, R., Jbilou, K., and Raydan, M. (2018). Conditional gradient method for double-convex fractional programming matrix problems. *Journal of Optimization Theory and Applications*, 176(1): 163–177.

Charnes, A., and Cooper, W. W. (1962). Programming with linear fractional functionals. *Naval Research Logistics Quarterly*, 9(3–4): 181–186.

Chergui, M. E. A., and Moulaï, M. (2008). An exact method for a discrete multiobjective linear fractional optimization. *Advances in Decision Sciences*, http://dx.doi.org/10.1155/2008/760191.

Das, S. K., and Mandal, T. (2015). Solving linear fractional programming problems using a new Homotopy Perturbation method. *Operations Research and Applications: An International Journal*, 3: 1–15.

Gupta, S., Raina, A. A., and Ali, I. (2017) An iterative algorithm for solving fractional linear programming problem. *International Journal of Recent Scientific Research*, 8: 17488–17493.

Hasan, M. B., and Acharjee, S. (2011). Solving LFP by converting it into a single LP. *International Journal of Operations Research*, 8: 1–14.

Jain, E., Dahiya, K., and Verma, V. (2018). Integer quadratic fractional programming problems with bounded variables. *Annals of Operations Research*, 269(1–2): 269–295.

Jiao, H., Liu, S., Yin, J., and Zhao, Y. (2016). Outcome space range reduction method for global optimization of sum of affine ratios problem. *Open Mathematics*, 14(1): 736–746.

Liu, X., Gao, Y. L., and Zhang, B., and Tian, F. P., (2019). A new global optimization algorithm for a class of linear fractional programming. *Mathematics*, 7(9): 867.

Martos, B. (1960). Hyperbolic Programming, Publications of the Research Institute for Mathematical Sciences. *Hungarian Academy of Sciences*, 5: 386–407.

Martos, B., and Whinston, V. (1964). Hyperbolic programming. *Naval Research Logistics Quarterly*, 11(2): 135–155.

Pandian, P. and Jayalakshmi, M. (2013). On solving linear fractional programming problems. *Modern Applied Science*, 7: 90.

Pramanik, S., Dey, P. P., and Giri, B. C. (2011). Multi-objective linear plus linear fractional programming problem based on taylor series approximation. *International Journal of Computer Applications*, 32: 61–68.
Saha, S. K., Hossain, M. R., Uddin, M. K., and Mondal, R. N. (2015). A new approach of solving linear fractional programming problem (LFP) by using computer algorithm. *Open Journal of Optimization*, 3: 74.
Shen, P. P., and Wang, C. F. (2006). Wang, Global optimization for sum of linear ratios problem with coefficients. *Applied Mathematics and Computation*, 176: 219–229.
Shen, P., Zhang, T., and Wang, C. (2017). Wang. Solving a class of generalized fractional programming problems using the feasibility of linear programs. *Journal of Inequalities and Applications*, 147. https://doi.org/10.1186/s13660-017-1420-1.
Simi, F. A., and Talukder, M. S. (2017). A new approach for solving linear fractional programming problems with duality concept. *Open Journal of Optimization*, 6(1): 1–10.
Sivri, M., Albayrak, I., and Temelcan, G. (2018). A novel approach for solving quadratic fractional programming problems. *Croatian Operational Research Review*, 9(2): 199.
Sun, X., Tang, L., Long, X. J., and Li, M., (2018). Some dual characterizations of Farkas-type results for fractional programming problems. *Optimization Letters*, 12: 1403–1420.
Tantawy, S. F. (2008a). A new procedure for solving linear fractional programming problems. *Mathematical and Computer Modelling*, 48: 969–973.
Tantawy, S. F. (2008b). An iterative method for solving linear fractional programming (LFP) problem with sensitivity analysis. *Mathematical and Computational Mathematics*, 13: 147–151.
Tantawy, S. F. (2014). A new concept of duality for linear fractional programming problems. *International Journal of Engineering and Innovative Technology*, 3: 147–149.
Zhang, Y. H., and Wang, C. F., (2017). A new branch and reduce approach for solving generalized linear fractional programming. *Engineering Letters*, 25: 14.

4 The Tradeoff in Managing Overall Cost and Backorder Minimization in Two-Stage Multi Commodity Supply Chain Problem

Vimal Kumar, Ajay Jha, Pratima Verma, and Nagendra Kumar Sharma

CONTENTS

4.1 INTRODUCTION

In India, food grains are distributed at subsidized rates to the poor people via a public distribution system which collects from surplus-producing states and distributes to other parts of country through the appropriate channels of distribution comprising of various warehouses (over 2000) and government-authorized shops. This challenging task is undertaken by the Food Corporation of India (FCI) with objectives to deliver food to the right place at the right quantity optimally, which often not is the case (Kumar, 2018). Generally, the grains are transported through railways using train rakes. The low-cost supply, optimized transportation, and

DOI: 10.1201/9781003089636-4

adequate buffers have important roles to play in the success of the supply chain process (Nagasawa et al., 2013; Seliaman, 2013). Though logistics optimization is an age-old problem, it is posing newer challenges in a globalized scenario. There are various problems, varying from single to multi-item and single to multi-period to get benefitted of the lot-sizing problem (Sharma et al., 2017). As far as the FCI problem is concerned, the warehouses in the surplus-producing states hold sufficient stock and often face backorders subject to limited availability of rakes for the specific location of demand points. The trucks and other road transport carriers are to be deployed in exigencies. The inventories have to be maintained at certain strategic intermediate points or the distribution points (customer's location) to meet the service criteria (Jawahar et al., 2012).

The current food grain supply chain design scenario uses an extended national framework where sources of supply are scattered at various locations nationwide with associated capacity and cost restrictions and the transportation process is done by road/rail. The problem also involves strategically locating intermediate points (warehouses or transshipment points) that can serve as pooling point for demands of various demand locations or distribution centers (DCs), so that an optimized rake/truckload is operated, thereby reducing the operational cost and time. Generally, there are fixed shipping costs with a variable component based on the number of units to be shipped on truck or rake with capacity constraints (Cachon, 2001). In addition, there are per unit warehouse space, holding and backorder penalty costs. Cost of purchase also depends upon the location and the number of items purchased. In brief, the source locations, transshipment points, and its planning is affected by two main criteria. The first one is transportation of goods cost and location of transshipment point (one time cost) along with considerations of cost of empty haulage of goods trains/truck and the delays in transporting the goods; and (ii) further planning is involved with estimating delays or backorders and inventory holdings to meet demands of distribution points. In the continuous shipments, sometimes shortages are intently carried out for optimized transportation. The distribution centers take decisions on the supply of backorders and lost sales occurs in case of various deficiencies (Nagasawa et al., 2013). The following research questions (RQs) are addressed in this work:

RQ1: *How the selection of warehouse location and transportation mode is ascertained with dispatching from plants to warehouses stage and later to distribution centers?*
RQ2: *How to identify, develop and address the model?*
RQ3: *How to deal with the trade-off between twin objectives of minimizing overall transportation-allocation cost and backorder cost?*

Generally, the service levels are compromised in aiming for overall cost optimization (Sarimveis et al., 2008). As such, no model illustration exists in the literature that highlights this tradeoff with restrictions of full load charge on transportation modes. The objective of this chapter is to model location-allocation capacity-constrained transshipment problem over a multi-period planning horizon where the transportation is charged on the full capacity. The model optimizes the overall cost and backorder minimization in a two-stage multi-commodity supply chain problem. We solve the model problem using a general algebraic modeling system (GAMS) for solving the mathematical optimization problem. This chapter is organized as follows: the literary background is given in section 4.2; the problem description and formulation with appropriate explanation have been discussed in section 4.3; computational results and discussion in section 4.4; finally, the conclusion and managerial insights in section 4.5.

4.2 LITERARY BACKGROUND

A supply chain process consist of a network handling a set of raw materials, semi-finished goods, and final products via supply, manufacturing and distribution activities (Voordijk, 1999). The supply chain distribution system reduces the overall transportation cost with different policies from the supplier to the consumer (Akbalik and Penzb, 2011). Swaminathan et al. (1998) considered two main categories of supply chain processes: the first one is the structural elements, which are used to model the production and transportation of products; the second is the control elements, which are used to specify various control policies related to the flow of information. Diaby (1991) considered two different types of the cost related to transportation problem: fixed and continuous charges. Silva and De La Figuera (2007) addressed the facility location problem that minimizes the overall transportation cost with the fixed cost. Demirli and Yimer (2008) and Xiao and Zheng (2010) highlighted the cost minimization in their supply chain modeling with the integration of supply chain process scattered at various locations and distribution centers. Francis et al. (1983) and Drezner (1995) tried to solve the capacitated facility location-allocation transshipment using NP-hard combinatorial problems. ReVelle and Eiselt (2005) described four elements: facilities; customers; space and metric for the location of facilities; and customer's indications. The variation of the facility location problem depends on the dynamic nature and structure of demand (Torres-Soto and Üster, 2011). Some studies believe that only new facilities can be installed during the particular planning horizon (Erlenkotter, 1981, Shulman, 1991; Drezner, 1995; Canel and Khumawala, 2001), whereas Roodman and Schwarz (1975) assume that they can only be closed due to changing, mainly decreasing, demand conditions. Geoffrion and Graves (1974) and Sharma et al. (2016) have solved a generalized multiple commodity version of the transshipment problem and have used the Benders decomposition to solve the 0–1 mixed-integer linear program. Sharma (1996) and Sharma et al. (2015) presented a MILP formulation for the single-stage ware location problem (SSWLP).

Our work models MILP location-allocation capacity-constrained transshipment problems over the multi-period planning horizon where the transportation is charged on full capacity.

Complete explanations of the abbreviations of indices, parameters, and decision variables used in this chapter:

4.2.1 Indices

t: Time period (t = 1.....4)
i: Plants locations (1 to I)
j: Warehouses at stage 1 (1 to J)
k: Distributors at stage 2 (1 to K)
l: Type of products (1, 2)
m: Mode of transportation (Road/Rail: =1, 2)

4.2.2 Parameters

$Cost_{lit}$: unit cost of product l at location i in time period 't'.
$Cost_{ljt}$: unit cost of product l at location j in time period 't'.
$Cost_{lkt}$: unit cost of product l at location k in time period 't'.
$CTrT_{ij}$: Cost of transporting one full truckload between plant location i and warehouse j
$CTrR_{ij}$: Cost of transporting one full rake load between plant location i and warehouse j
$CTrT_{jk}$: Cost of transporting one full truckload between warehouse location j and D.C. k
$CTrR_{jk}$: Cost of transporting one full rake load between warehouse location j and D.C. k
r_1: rate of interest for holding inventory at location j for the planning period T.
r_2: rate of interest for holding inventory at location k for the planning period T.
r_b: rate of interest to account for the unit backorder at location k for the time period t.
$C_{Space}WS1_j$: fixed cost of the warehouse place space step 1 at another place 'j' amortized for planning period T.
$CapWS1_{jl}$: Capacity of the warehouse at location 'j' of commodity l.
$CapDC_{jl}$: Capacity of warehouse/D.C. at location 'k' of commodity l.
D_DC_{tlk}: Demand at D.C. step 2 in time period 't' of commodity l at place 'k'.
$Supply_{itl}$: The Supply capacity of commodity l at plant location i at time t.
V_l: Volume/weight of commodity l.
$CapTr_{ml}$: Capacity of full truck/Rake for commodity l of transportation mode m.

4.2.3 Decision Variables

$XPWS_{lijt}$: Quantity of product l moved from plant i to warehouse j at stage 1 in time period 't'.
$XWSDC_{ljkt}$: Quantity of product l moved from warehouse j to distribution center k at stage 2 in time period 't'.
XDC_BO_{tk}: Backorder quantity of commodity l at distribution center k in time period 't'.
Inv_WS1_{ljt}: Inventory of product l at warehouse j stage 1 at time period 't'.
Inv_DC_{lkt}: Inventory of product l at distribution center k stage 2 at time period 't'.

$NTWS_{ijt}$: Number of trucks loads transported from plant location i to distribution center j

$NRWS_{ijt}$: Number of trains loads transported from plant location i to distribution center j

$NTDC_{jkt}$: Number of trucks loads transported from warehouse location j to distribution center k

$NRDC_{ijt}$: Number of trains loads transported from warehouse location j to distribution center k

y_j: Decision variable for locating a warehouse at location j {value =1 if located, else =0}

4.3 PROBLEM DESCRIPTION AND FORMULATION

We consider multi-commodity multi-market (distribution centers) with a known rough estimate of demand per quartile; where demand is met through distant source points via intermediate transshipment points or regional warehouses. Thus, the food grains supplied to the transshipment points/regional warehouses are further delivered to the distribution centers through an optimized plan. Our model therefore not only locates the warehouse points, but also allocates the distribution centers to these points. The problem involves two-stage management of shipment of goods/orders: first from various supply points to intermediate warehouses and second from these intermediate points to demand centers. For each order shipped, we use a carrier with a fixed capacity, such as a truck and train/rake whichever found necessary according to the ordered quantity and the capacity of the carrier. There is a constraint on charging full truck or full rake load, even if the shipment is a partial truck or partial rake load. We have eliminated the complexity of braking tariff into fixed and variable components by this assumption. However, full truck/rake charge constraint often results in the delaying of orders until full load demand is achieved, often resulting in shortages or backorders. Shortage often implies lost sales whereas backorder refers to fulfillment of order in the next period (often incurring some rebate or incentive to the customer for waiting). In this present research, we focus on the overall cost and backorder minimization in the two-stage multi-commodity supply chain problem. Figure 4.1 represents the supply chain models of the network for multi-products, plants, warehouse, and distribution centers.

We focus on cost structures, which consist of the cost of transportation, cost of purchase, inventory carrying cost, backorder cost, and cost of warehouse space, as commonly observed in supply chain design. The method is based on the presence of collective production of unit. (Jiang, 2019). Overall cost and backorder minimization problems are to be determined over the planning horizon (say 't' periods) for multi-commodity items. Probable demand for each item (k index for items = 1, 2...K) for the given period is denoted as D_DC_{tlk} with time period t = t1, t2, t3, t4 known a priori, and must be satisfied from current supply and inventory or by backorder.

$XPWS1_{itj}$ is quantity or the amount placed from the supply point to the fixed warehouse step 1 that is rightly associated to the railway "*broad gauge*" connectivity. $XWS1DC2_{itjk}$ is quantity or the amount placed from the fixed warehouse step 1 to Distribution Center step 2, which can be linked to both road and railway. A Mathematical Model has been developed for distribution of food grain.

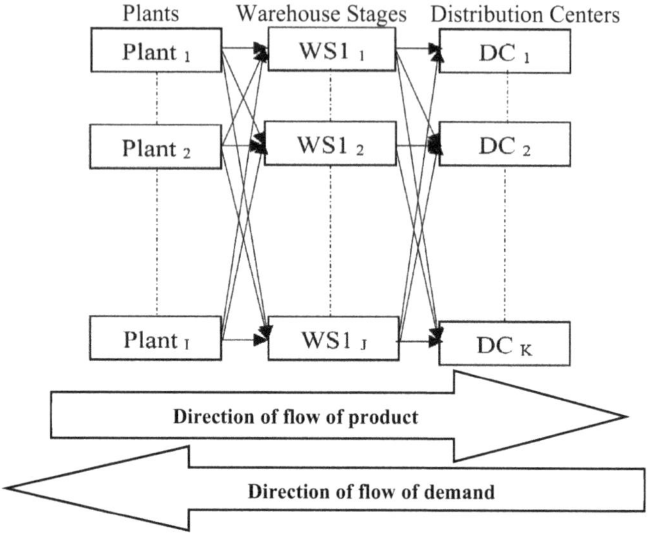

FIGURE 4.1 The network for multi-products, plants, warehouse, and the distribution centers.

We have taken some assumptions, which are as follows: (i) The mode of dispatch of these commodities to distribution centers is generally by road, rail or water, but here we consider only the rail and road modes of transportation. (ii) Warehouse sizes are unlimited. (iii) Cost includes transportation; purchase; inventory carrying; BO; and warehouse space. (iv) The cost of the truck/rake waiting time at plants, WHs, and DCs is zero. (v) Here, we avoid (like some FMCG type) products stock out, but it results in excessive inventory.

4.3.1 Mathematical Formulation

We have divided our planning horizon into four demand seasons or time periods (t = 1 to 4); total possible plant or supply locations are I (here food grain collection points); possible warehouse locations or transshipment points at stage 1 as J, and known distribution or demand points as K or distributors at stage 2. In our model, we have taken just two types of products (I = 1, 2), but more can be taken by extending the product index limit from 2 to higher number and including its demand values at Distribution Centers (D.C.s). Similarly, only two transportation modes are taken in our model, rail and road (m = 1, 2), which can be extended to more with the change in the respective parameter. We define the parameters first, and then decision variables and finally our model. The explanation of parameters is given at the start of chapter.

Formulation based on Mathematics

The Objective function 1

Minimize (Cost of transportation + Cost of Purchase + Cost of inventory carrying + B.O. cost + Cost of a the fixed warehouse space at certain location)

$$\sum_{i,j,t} \mathbf{NTWS_{i,j,t}CTrT_{i,j}} + \sum_{i,j,t} \mathbf{NRWS_{i,j,t}CTrR_{i,j}}$$
$$+ \sum_{j,k,t} \mathbf{NTDC_{j,k,t}CTrT_{j,k}} + \sum_{j,k,t} \mathbf{NRDC_{j,k,t}CTrR_{j,k}}$$
$$+ \sum_{l,i,j,t} \mathbf{XPWS_{l,i,j,t}Cost_{l,i,t}} + \sum_{l,j,t} \mathbf{XWS\_INV_{ijt}Cost_{l,j,t}\, r_1}$$
$$+ \sum_{l,k,t} \mathbf{XDC\_INV_{l,k,t}Cost_{l,k,t}\, r_{21)tS1ng\ number\ of\ e\ full\ truck\ load\ between\ nt\ location\ i\ ng\ period\ T.}}$$
$$+ \sum_{l,k,t} \mathbf{XDC\_BO_{lkt}Cost_{l,k,t}\, r_{31)tS1ng\ number\ of\ e\ full\ truck\ load\ between\ nt\ location\ i\ ng\ period\ T.}}$$
$$+ \sum_{j} \mathbf{y_jCSpaceWS_j}$$

Objective function 2
Minimizing the number of Backorders

$$\mathbf{Min.} \sum_{lkt} \{\mathbf{XDC\_BO_{lkt}}\}$$

Subject to

$$\sum_{i} XPWS_{l,i,j,t} + Inv\_WS_{l,j,t-1} = \sum_{k} XWSDC_{ljkt} + Inv\_WS_{l,j,t} \quad \forall l, j, t \tag{4.1}$$

$$\sum_{j} XWSDC_{l,j,k,t} + Inv\_DC_{l,k,t-1} = D_{l,k,t} + Inv\_DC_{l,k,t} \quad \forall l, k, t \tag{4.2}$$

$$\sum_{j} XPWS_{lijt} \le Supply_{lit} \quad \forall l, i, t \tag{4.3}$$

$$\sum_{i} XPWS_{lijt} \le CapWS1_{lJ} \quad \forall l, j, t \tag{4.4}$$

$$\sum_{j} XWSDC_{ljkt} \le CapWS2_{lk} \quad \forall l, k, t \tag{4.5}$$

$$XPWS1_{lijt} + XPWS2_{lijt} = XPWS_{lijt} \tag{4.6}$$

$$XWSDC1_{lijt} + XWSDC2_{lijt} = XWSDC_{lijt} \tag{4.7}$$

$$NTWS_{ijt} = Int\left[\sum_{l} \{XPWS1_{lijt} * V_l\} / CapTr_{l1}\right] \quad \forall i, j, t \tag{4.8}$$

$$NRWS_{ijt} = \text{Int}\left[\sum_{l}\{XPWS2_{lijt} * V_l\} / CapTr_{21}\right] \quad \forall i, j, t \tag{4.9}$$

$$NTDC_{jkt} = \text{Int}\left[\sum_{l}\{XWSDC1_{ljkt} * V_l\} / CapTr_{11}\right] \quad \forall j, k, t \tag{4.10}$$

$$NRDC_{jkt} = \text{Int}\left[\sum_{l}\{XWSDC2_{ljkt} * V_l\} / CapTr_{21}\right] \quad \forall j, k, t \tag{4.11}$$

$$XWS\_INV_{ljt} = \max\{Inv_{WSljt}, O\} \quad \forall l, j, t \tag{4.12}$$

$$XWS\_INV_{lkt} = \max\{Inv_{WSlkt}, O\} \quad \forall l, k, t \tag{4.13}$$

$$XDC\_BO_{lkt} = -\min\{Inv_{DClkt}, O\} \quad \forall l, k, t \tag{4.14}$$

$$\sum_{lit} XPWS_{lijt} \le y(j) * M \quad \forall j \tag{4.15}$$

$$\sum_{lkt} XWDCS_{ljkt} \le y(j) * M \quad \forall j \tag{4.16}$$

$$XPWS_{l,i,j,t} \ge 0 \quad \forall l, i, j, t \tag{4.17}$$

$$XWSDC_{l,j,K,t} \ge 0 \quad \forall l, j, k, t \tag{4.18}$$

$$y_j \in \{0,1\} \tag{4.19}$$

The objective function one is considered as the sum of transportation cost (by using truck and train/rake), cost of purchase, inventory-carrying cost, BO cost, and the cost of locating a warehouse. The objective function two is considered as the number of BO.

Equation (4.1) indicates flow balance as the sum of quantity shipped from all the plants to each warehouse for each commodity in each time period and previous period inventory is equal to the sum of quantity shipped from each WS to all DCs for the same commodity in the same time period and current period inventory. Equation (4.2) indicates that that sum of quantity shipped from all the WS to each DC for each commodity in each time period and previous period inventory is less than or equal to the sum of demand at each DC for the same commodity in the same time period and current period inventory. Equation (4.3) indicates that sum of quantity shipped from each plant to all the warehouses for each commodity in each time period is less than or equal to the supply capacity of the same plant for the same commodity in the same time period. Equation (4.4) indicates that the sum of quantity shipped from all the plants to each warehouse for each commodity in each time period is less than or equal to the capacity of the same WS for the same commodity in the same time period. Equation (4.5) indicates that sum of quantity shipped from all the WS to each DC for each commodity in each time period is less than or equal to the capacity of the same DC for the same commodity in the same time period.

Equations (4.6) to (4.11) give the model, freedom of choosing the desired mode of transportation (here train/rake and truck) in thre optimum number. Therefore, the quantities to be shipped are bi-split as here only two modes of transportation are considered.

Equations (4.12) to (4.14) take care of the inventory and backorder costs. Inventory carrying cost is considered when inventory is positive. Negative inventory at DC will be treated as a backorder and will be described as the backorder cost. Almost every time backorder cost is greater than the inventory carrying the cost. Equations (4.15) and (4.16) are weak linking constraint, where y(j) is the binary decision variable and M is a very large number. Equations (4.17) and (4.18) are non-negativity constraints on the decision variables while Equation (4.19) is a binary constraint on the WS location variable y(j).

4.4 COMPUTATIONAL RESULTS AND MANAGERIAL INSIGHTS

The problems are generated by assigning range of values to supply capacity, warehouse locations and demand points. The values are randomized for 30 sets of problem data in the pre-determined range. The results are obtained using GAMS software with using MILP model given there. Table 4.1 and Table 4.2 give the computational results of the above formulations while Figure 4.2 and Figure 4.3 give the total cost and total backorder for the problem size thirty respectively.

Looking at the range of values obtained from the two data sets we see when the primacy is given to the minimization of overall total cost for warehouse location and transportation objective in meeting the demand, the number of backorders increases in value and when the number of backorders decreases then the total supply chain cost increases. When the primacy is given to avoid the backorders or fulfilling the demand at priority, the model permits the partially loaded transport by the rakes and trucks. So the model provides a tradeoff in overall cost minimization and the service level. In addition, the volume consideration taken in account for the two commodities limits the space utilization. The formulation could be solved as two separate single objective problems. The first case illustrates the number of rail rakes necessary at a specific time and the number of trucks required, as well as the choice of the transshipment point to serve a particular demand point. The limits can be imposed on the number of backorders permitted or assigning a high penalty cost to the backorders. The bi-objective formulation gives the clear tradeoff in handling the back orders and overall supply chain cost.

TABLE 4.1
Result Table for Cost (30 Problems for Objective Function 1 (Z1) - Cost Minimization and Objective Function 2 (Z2) - Backorder Minimization)

Cost		
Problem Number	Objective Function 1 (Z1) - Cost Minimization	Objective Function 2 (Z2) - Backorder Minimization
Problem 1	39718738.52	45106833.47
Problem 2	40211554.89	44982250.32
Problem 3	39173932.83	44453110.59
Problem 4	39451428.99	44451348.53
Problem 5	39645305.39	44794545.87
Problem 6	38667847.62	44414354.55
Problem 7	40523782.41	45809775.45
Problem 8	39119241.86	44187932.59
Problem 9	39802005.62	45398200.43
Problem 10	40084344.37	45453124.79
Problem 11	38752595.79	44430754.49
Problem 12	39839422.05	44859099.2
Problem 13	39323916.35	45181023.6
Problem 14	39972568.93	45117058.13
Problem 15	40714633.24	46378084.2
Problem 16	40306923.28	45486113.47
Problem 17	39187209.64	44452597.71
Problem 18	40112328.33	45626851.55
Problem 19	40106555.41	45753178.9
Problem 20	39174923.39	44246380.29
Problem 21	39646376.91	44517723.84
Problem 22	39196134.11	44401800.34
Problem 23	40061809.69	45704605.66
Problem 24	40574840.49	45980731.37
Problem 25	40415904.96	45164405.7
Problem 26	39440256.57	44427529.73
Problem 27	40804173.09	45626023.86
Problem 28	40130280.94	45272748.6
Problem 29	39125770.32	44252142.36
Problem 30	40125298.97	45571912.72

TABLE 4.2
Result Table for Backorder (30 Problems for Objective Function 1 (Z1) - Cost Minimization and Objective Function 2 (Z2) - Backorder Minimization)

Backorders		
Problem Number	Objective Function 1 (Z1) - Cost Minimization	Objective Function 2 (Z2) - Backorder Minimization
Problem 1	102117.2373	3757.761296
Problem 2	106889.6981	11385.78431
Problem 3	96002.56615	2170.818368

(Continued)

TABLE 4.2 (Continued)

Backorders

Problem Number	Objective Function 1 (Z1) - Cost Minimization	Objective Function 2 (Z2) - Backorder Minimization
Problem 4	105209.6172	12845.30755
Problem 5	101973.0585	8010.05639
Problem 6	99988.12397	3651.47725
Problem 7	105291.1555	6288.13113
Problem 8	98417.27751	2537.93179
Problem 9	100468.5914	4383.69587
Problem 10	104718.6204	8246.914186
Problem 11	100262.7759	5407.93242
Problem 12	114467.8196	8366.902426
Problem 13	103837.9924	7804.078126
Problem 14	101929.7468	7590.452426
Problem 15	109365.9634	11767.4692
Problem 16	103297.1945	10440.7994
Problem 17	101073.0195	6629.551232
Problem 18	101785.8679	4376.723854
Problem 19	101533.7267	7817.871462
Problem 20	98686.01163	4178.837214
Problem 21	97895.86569	6282.960826
Problem 22	100369.9192	5715.642144
Problem 23	103135.9432	6589.700888
Problem 24	104049.0799	4990.131156
Problem 25	129441.4205	7974.532174
Problem 26	104646.7432	9581.40728
Problem 27	98937.21324	1923.186522
Problem 28	102103.7477	7664.623128
Problem 29	96450.21067	1118.987664
Problem 30	107886.5841	12478.32476

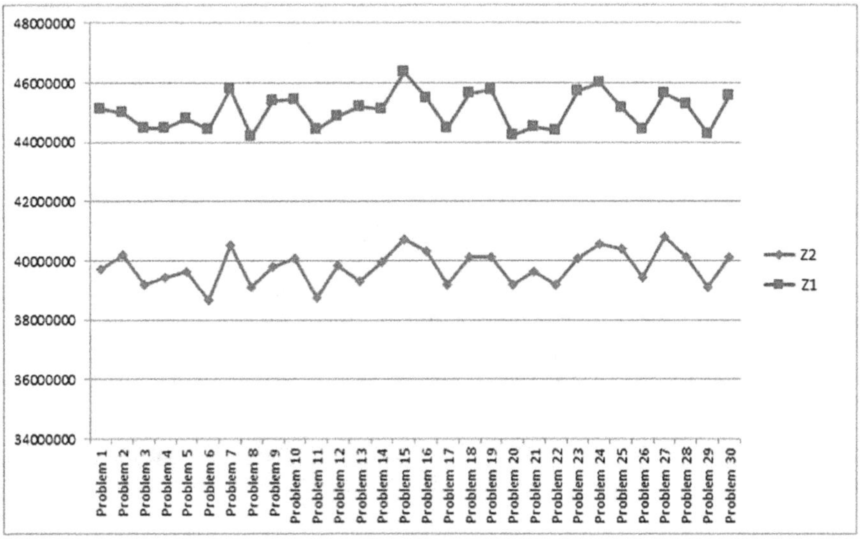

FIGURE 4.2 Total cost for the problem size 30.

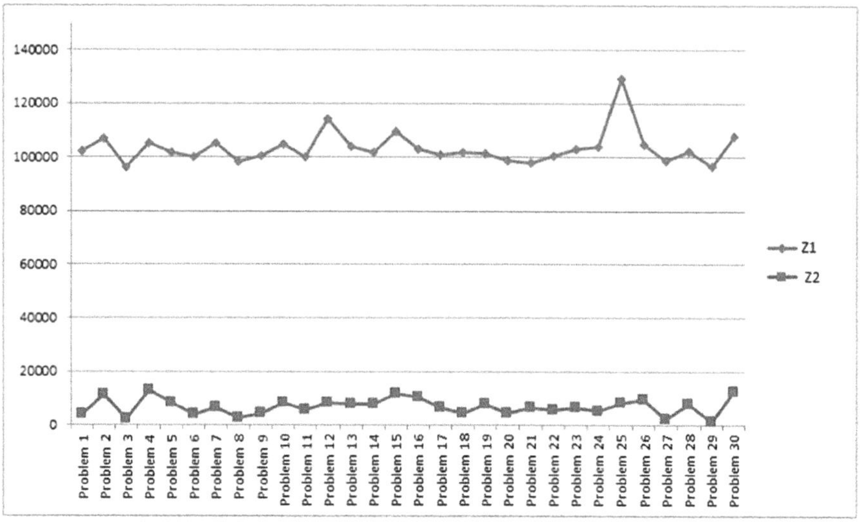

FIGURE 4.3 Total backorder for problem size 30.

4.5 CONCLUSION

This study examines the interdependence between selections of intermediate transshipment/warehouse locations, transportation modes and backorders to serve a given set of demand sites. The MINLP model so developed also helps in ascertaining the quantity of a particular product to be transported from supply points to each transshipment/warehouse and further to distribution centers. It optimizes the total transportation, purchase, backorder, inventory carrying, and warehouse space costs, as well as the backorder level. The supply chain optimization model so developed is useful for organizations like the Food Corporation of India (FCI), where food grains are collected from surplus states and stored at various locations to further transport to various distribution/demand centers through multiple modes (train/trucks). The GAMS software is used to optimize the twin objectives of total supply chain cost minimization and the minimization of backorders under the capacity-constrained supply and transshipment points, and the transportation cost is charged on the full truck or full rake load, even when the load is only partial. The model provides the location of transshipment points, dispatch orders along with the requirements of railway rakes/wagons or trucks from the collection centers (plants) to the warehouse at stage one while dispatching from warehouses stage one to distribution centers at stage two. A tradeoff between these twin objectives is observed indicating compromise of overall optimization of cost and back order with rising service levels.

The model thus developed will assist the decision-makers in taking a proactive approach before the demand realization period in maintaining the transportation system network along with transshipment point planning to meet the diverse demand needs. The managers of a manufacturing organization can utilize this model for transportation and warehouse location problem simultaneously, as well as keeping the service level in consideration. Efficiently achieving cost-effective distribution along with an adequate service level is the benefit of this model. In addition, incorporating different shipment policies and more echelons of supply chain in the study are another potential and interesting area for future work.

REFERENCES

Akbalik, A., & Penz, B. (2011). Comparison of just-in-time and time window delivery policies for a single-item capacitated lot sizing problem. *International Journal of Production Research*, 49(9), 2567–2585.

Cachon, G. (2001). Managing a retailer's shelf space, inventory, and transportation. *Manufacturing & Service Operations Management*, 3(3), 211–229.

Canel, C., & Khumawala, B.M. (2001). International facilities location: A heuristic procedure for the dynamic uncapacitated problem. *International Journal of Production Research*, 39(17), 3975–4000.

Demirli, K., & Yimer, A.D. (2008). Fuzzy scheduling of a build-to-order supply chain. *International Journal of Production Research*, 46(14), 3931–3958.

Diaby, M. (1991). Successive linear approximation procedure for generalized fixed-charge transportation problems. *Journal of the Operational Research Society*, 42(11), 991–1001.

Drezner, Z. (Ed.). (1995). *Facility location: A survey of applications and methods*. Springer Verlag.

Erlenkotter, D. (1981). A comparative study of approaches to dynamic location problems. *European Journal of Operational Research*, 6(2), 133–143.

Francis, R.L., McGinnis, L.F., & White, J.A. (1983). Locational analysis. *European Journal of Operational Research*, 12(3), 220–252.

Geoffrion, A.M., & Graves, G.W. (1974). Multicommodity distribution system design by Benders decomposition. *Management Science*, 20(5), 822–844.

Jawahar, N., Gunasekaran, A., & Balaji, N. (2012). A simulated annealing algorithm to the multi-period fixed charge distribution problem associated with backorder and inventory. *International Journal of Production Research*, 50(9), 2533–2554.

Jiang, Z. (2019). "Aggregate planning", production and operation managements McGraw-Hill/Irwin chapter 3, retrieved https://slideplayer.com/slide/8811196/ on 20th sept. 2019.

Kumar, V. (2018). *Equal distribution of shortages in supply chain problem of food corporation of India. Eighth International Conference on Industrial Engineering and Operations Management (IEOM) held in Bandung, Indonesia*, March 6–8, 2018, pp. 1972–1984. IEOM Society International. http://ieomsociety.org/ieom2018/papers/532.pdf

Nagasawa, K., Irohara, T., Matoba, Y., & Liu, S. (2013). Joint replenishment problem in multi-item inventory control with carrier capacity and receiving inspection cost. *Operations and Supply Chain Management: An International Journal (OSCM)*, 6, 111–116.

ReVelle, C.S., & Eiselt, H.A. (2005). Location analysis: A synthesis and survey. *European Journal of Operational Research*, 165(1), 1–19.

Roodman, G.M., & Schwarz, L.B. (1975). Optimal and heuristic facility phase-out strategies. *AIIE Transactions*, 7(2), 177–184.

Sarimveis, H., Patrinos, P., Tarantilis, C.D., & Kiranoudis, C.T. (2008). Dynamic modeling and control of supply chain systems: A review. *Computers & Operations Research*, 35(11), 3530–3561.

Seliaman, M.E. (2013). Optimizing the two-stage supply chain inventory model with full information sharing and two backorders costs using hybrid geometric-algebraic method. *Journal of Optimization*, 2013, 1–5.

Sharma, R.R.K. (1996). Food grains distribution in the Indian context: An operational study. *Operations Research for Development, Ahmedabad, India*, 212–227.

Sharma, R.R.K., Kumar, V. and Khan, N.D. (2017) Developing a new reformulation of single level capacitated lot sizing problem (SLCLSP) with set up, shortage and inventory costs. *American Journal of Operations Research*, 7(5), 272–281.

Sharma, R.R.K., Malviya, A., Kumar, V., Singh, V. and Agarwal, P. (2016). Application of modified benders decomposition to single-stage multi-commodity multi-period warehouse location problem: An empirical investigation. *American Journal of Operations Research*, 6(3), 245–259.

Sharma, R.R.K., Tyagi, P., Kumar, V. & Jha, A. (2015). Developing strong and hybrid formulation for the single stage single period multi commodity warehouse location problem: Theoretical framework and empirical investigation. *American Journal of Operations Research*, 5(3), 112–128.

Shulman, A. (1991). An algorithm for solving dynamic capacitated plant location problems with discrete expansion sizes. *Operations Research*, 39(3), 423–436.

Silva, F.J.F., & De la Figuera, D.S. (2007). A capacitated facility location problem with constrained backlogging probabilities. *International Journal of Production Research*, 45(21), 5117–5134.

Swaminathan, J.M., Smith, S.F., & Sadeh, N.M. (1998). Modeling supply chain dynamics: A multiagent approach. *Decision Sciences*, 29(3), 607–632.

Torres-Soto, J.E., & Üster, H. (2011). Dynamic-demand capacitated facility location problems with and without relocation. *International Journal of Production Research*, 49(13), 3979–4005.

Voordijk, H. (1999). Obstacles and preconditions for logistics and manufacturing improvements in Africa–a case study. *International Journal of Operations & Production Management*, 19(3), 293–307.

Xiao, J., & Zheng, L. (2010). A correlated storage location assignment problem in a single-block-multi-aisles warehouse considering BOM information. *International Journal of Production Research*, 48(5), 1321–1338.

5 An Optimization Study on Behavior of Actinide Monochalcogenides at High Pressure

Ritu Saxena, Neetu Paliwal, Pankaj Kumar Jha, and Ramakant Bhardwaj

CONTENTS

5.1 INTRODUCTION

Rare earths are the subject of increasing study because of their several interesting electronic, mechanical, thermal and structural properties. The rare earth actinide compounds have proved of particular interest. Among the series of these rare earth actinides, plutonium monochalcogenides PuX; X = S and Se are shown to exhibit anomalous behaviour because of their lattice constants. Comparative studies of plutonium monochalcogenides have been carried out. This investigation confirms the inadequacy of spin-polarized density functional theory (DFT). Plutonium compounds are found to be paramagnetic in experiments at low temperatures [1–5]. At atmospheric pressure, these PuS and PuSe are in $B_{1\,phase}$. With an increase in pressure, they transform into CsCl structure.

On the basis of electronic structure calculation, the anomalous properties of PuX studied [6]. Hasegawa et al. studied the various properties of PuX [7]. These investigations suggest that these compounds are semi-metallic in nature. Other studies are

DOI: 10.1201/9781003089636-5

TABLE 5.1
Generated Model Parameters

Compound	B	ρ	f (r)
PuS	31.1784	0.254	0.01271
PuSe	35.9621	0.267	0.01314

also carried out in divalent and trivalent states of PuX [8, 9]. They conclude the relativistic semiconductor behaviour of PuX. Magnetic state, as well as the electronic structure of plutonium from "first principles" calculations have been investigated. Shrivastava et al. [10] also studied for the present group of compounds. In this work, (TBIP) approach has been used. Also related work of optimize study on phase transition has done in [11–21]. Generated model parameters can be viewed in Table 5.1 for detailed study.

5.2 BRIEF ABOUT OPTIMIZATION IN PHASE TRANSITION STUDY

A given assembly of atoms (or molecules) may be either homogeneous or non-homogeneous. The homogeneous parts of such an assembly, called phases, are characterized by the thermodynamic properties such as pressure, temperature volume and energy. An isolated phase is stable only when its energy, or more generally its free energy, is minimum for the specified thermodynamic conditions energy barriers, the system is then said to be in a metastable state. If barriers do not exist, the state of the system becomes unstable and the system moves into a stable or equilibrium state characterized by the lowest possible free energy. As the temperature, pressure or any other variable such as an electric or magnetic field acting on a system is varied, the free energy of the system changes smoothly and continuously. Whenever such variations of the free energy are associated with changes in structural details of the phase (atomic or electronic configuration), a phase transformation (a phase transition) is said to occur. In this chapter we are mainly concerned with the study of the high-pressure phase transition in the alkaline earth chalcogenides of rock salt structure. Macroscopic as well as microscopic properties of the system are influenced by the transitions. Consequently, below the transition temperature or pressure they acquire distinct properties such as spontaneous polarization, magnetization, nonlinearities in their electromagnetic behaviour etc. Several properties of the system sometimes attain large changes in their magnitude near T_c and P_t. These changes are normally known as anomalies. Anomalous change in elastic coefficients, dielectric constants expressivities and specific heat is commonly observed.

A substance is usually present in one of three states: solid, liquid or gaseous state. Each of these states is called a phase. A phase is defined as a system, which is homogeneous having defined boundaries. It may be a chemically pure substance or may have more than one component, such as air or a mixture of two miscible liquids.

Phase transition is a process in which a thermodynamic system changes from one state to another with different physical properties, solid, liquid and gas. The transition from solid to liquid phase is marked by a discontinuous change in shearing strength; a transition to gaseous phase is attended with discontinuous change in density and so on. In most cases, phase transition takes place by appropriate changes in temperature and/or pressure. The melting of ice into water and evaporation of water into vapor are common examples of phase transitions and are satisfactorily explained on the basis of equation of state.

All phase transitions are essentially due to the interplay of attractive forces between the particles of the system and their thermal motion. While attractive forces tend to hold them together and trap them in bound states, thermal energy leads to random and free motion. If insufficient thermal energy is present, particles get trapped as in plasma-gas, gas-liquid and liquid-solid phase transitions. The reverse transitions occur when enough thermal energy is available. Such transitions are called first-order transitions.

There are several types of phase transition, some of which are discussed below:

5.2.1 Reconstructive Type

In the process of the reconstructive phase transition the original linkages of the net are disrupted and the constituents of the solid reconstruct a new lattice; a typical example of such structural phase transition (SPT) is the transformation of aragonite to calcite in $CaCO_3$ at 723K. Due to inherent matter transport, these transitions are often slow, this is even more the case because of discontinuous symmetry change and it is considered to be of the first order.

5.2.2 Martensitic Type

Martensitic-Type transitions have certain characteristic structural features observed in both metallic and nonmetallic systems. Such transitions occur through the sharing of discrete volumes of material without changing the chemical compositions. The transitions often occur with extremely high velocities; they exhibit large temperature hysteresis. These are reversible and generally polymorphic in nature. One example is Fe-29% Ni alloys, in which martensitic transition takes place at –30C.

5.2.3 Distortive Type

In distortive phase transition, the linkages of the net are not disrupted, but the lattice is only slightly distorted. These transitions can further be classified into three categories, viz the electronically induced type, the displacive type and the order disorder type.

The energy of excitation in certain systems couples with the phonons, leading to a possibility of a change that can be introduced by the electronic excitations of certain constituents. If the coupling is a sensitive function of temperature, the transition may be triggered with change in the temperature and are referred as electronically induced transitions, the cooperative Jhon-Teller transition Band Jhon-Teller transition are few such transitions that fall in this class.

If the distortive type phase can be described through the super position of the atomic displacement occurring through longitudinal/transverse as well as optic/acoustic mode, the transition is known as displacive. This transition is classified in various subcategories. An order disorder transition take place due to long range ordering of atoms/molecules/ions. The distinction between order disorder and displacive transitions can be understood in terms of single cell potential $V(Q) = aQ^2 + bQ^4$ here, a, 0 and b. 0 and Q refers to generalized coordinate, which could stand for the displacive motion of atoms, orientation of molecule etc. The nature of Q specifies the physical nature of transition and helps in identifying the different types of order-disorder and displacive transitions.

In positional order-disorder transitions ordered state refers to situation in which atoms/molecules occupy either of the two positions selectively (position 1 or 2), while the disordered state arises when atoms occupy position 1 and 2 randomly. The examples of positional order–disorder transitions are AgI, CuZn, etc. In orientational order–disorder transition Q is associated with the orientational dynamics of molecular units, such as CN in KCN [18], NO_2 in $NaNO_2$, and NH_4 in NH_4Cl. One can also identify Q with spin orientation in magnetic transitions and electric dipoles in ferroelectric and antiferroelectric transitions.

Cooperative displacement of atoms in displacive limit occurs in three distinct ways, leading to ferrodistortive type. On the one hand, an optical phonon of long wavelength (i.e. q>0), the transition is ferrodistortive type. On the other hand, an optical phonon at Brillion Zone boundary results in an antiferrodistortive-type transition. In thermoelastic transition, Q is an acoustic phonon of long wavelength. Phase transition $BaTiO_3$, $SrTiO_3$ and $TeO_2$are, respectively, the physical examples of above mentioned transitions.

The first-order phase transition involves a sharp and discrete change in the internal energy of the system at T_c or P_t, which is the origin of discontinuity in several physical properties of the system. The transition exhibits non-zero heat of transition. The order parameter and the first derivative of Gibbs free energy (G) with respect to the order parameter also changes discontinuously. At the transition point, the two states of the system do not change at, while the heat capacity, thermal expansivity and compressibility Tc undergo discontinuous changes. The second derivative of G also goes through discontinuous change at Tc. The order parameter increases monotonically from zero (at Tc) towards a certain magnitude on lowering the temperature (Figure 3.1 (a and b)). Two states of the system tend to become identical as the transition point is reached, implying that the symmetry of the system at Tc contains all the elements of all the phases. The temperature defining the limits of the two phases as well as with Tc. The symmetry of the low temperature phase is the subgroup of the symmetry of the high temperature phase.

5.2.4 Structural Phase Transitions

When a solid undergoes phase transition by absorbing thermal energy, the transferred phase will possess higher internal energy. In the high temperature phase, the bonding between neighbouring atoms or units would be weaker than during the low temperature phase. This results in a change in the nature of the first nearest neighbours (primary coordinates) or the next nearest neighbours (secondary coordinates). Burger has classified phase transitions on the basis of the structural changes involving primary or higher coordinates. He also relates the potential barriers for the transition (or the transition speed) to structural changes. The transformations, where there are changes in primary coordination, will involve more drastic changes in energy as compared to those where only second or higher coordination changes. This can be easily visualized in the case of ionic crystal, where energies of the ionic bonds vary inversely as the interionic separation, while the energies of the Vander walls interaction vary as the sixth power of such separations.

Burger had defined transformations of the bond type, where two polymorphs differ greatly in terms of the nature of bonding. Examples of this kind are polymorphs of tin (grey and white), carbon (diamond and graphite) and phosphorus (yellow and black). It is well recognized today that the knowledge of crystal chemistry in terms of atomic arrangements of and bonding in crystals is essential for understanding the properties of solids and thus, it can provide a basis for classification and prediction of the nature of phase transitions. In making such a prediction we take into account that thermal phase transition generally occurs from structure of low symmetry (and high order). A positive volume change also accompanies the transition of the first order. This relates to the fact that the high temperature phase tends to have greater openness structure and lower coordination. While an application of pressure will facilitate a transition involving an increase in coordination (negative volume change).

Let us consider cubic AB-type compounds. The compounds can have NaCl, CsCl or ZnS structure with 6:6, 8:8 or 4:4 coordination respectively, depending on their radius ratios. We can predict the following types of transitions in AB compounds:

1. NaCl structure $\xrightarrow{\text{Pressure}}$ CsCl structure

   Examples: NaCl, KCl, RbCl, NaBr, KBr, monoxides and monosulfides of some metals.
2. CsCl structure $\xrightarrow{\text{Heat}}$ NaCl structure

   Examples: CsCl, NH_4Cl, NH_4Br, etc.
3. ZnS structure $\xrightarrow{\text{Pressure}}$ NaCl structure

   Examples: AgI, NH_4F, II-VI, IV-VI and III-V semiconductors and their alloys.
4. NaCl structure $\xrightarrow{\text{Heat}}$ Wurtzite structure

   Examples: MnS.
5. ZnS structure $\xrightarrow{\text{Heat}}$ B23 structure

   Examples: AgI.
6. NaCl structure $\xrightarrow{\text{Pressure}}$ B9 structure

   Examples: AgCl.
7. CsCl structure $\xrightarrow{\qquad}$ distorted NaCl structure

   Examples: $RbNo_3$.

In the simplest ionic crystal, it is possible to employ Born theory to predict the nature of the phase transitions. According to Born criterial in rigid ion model the relative stability can be accounted for in terms of the cation anion radius ratio. For radius ratios larger than 0.72, the CsCl structure is preferred and still lower values of the radius ratios favor the ZnS structure. The simplicity of the Born model tempted workers to modify it marginally and account for the phase stability. It is thus important to note that there are two main assumptions built into the Born model:

- Ions are point-like changes and
- The interionic forces are central in character.

Both these assumptions cause serious difficulties when small energy differences between structures are to be evaluated. Also, in many of the ionic solids covalent contributions are not negligible and their presence seriously limits the validity of the Born model. Later on, a theory of covalency and iconicity in crystals has been developed by Philips, which explains a number of properties in $A^{N}B^{8\text{-}N}$ crystals satisfactorily.

The present three-body potential discussed above has been used to study the structural and elastic behaviour of some ionic crystals. It is well known that the polymorphic phase transition is one of the important phase transition in solids is a cross-disciplinary subject of vital current interest. There exists a huge literature on this subject, which abounds both theoretical and experimental investigations of pressure-induced phase transition in several systems of solids. In the present investigation, we are, however, interested in the pressure-induced polymorphic phase transition in ionic solids. The important ionic solids crystallize in the rocksalt, zinc blende or cesium chloride structures. It is experimentally known that the application of a suitable hydrostatic pressure to the crystals originally in the rocksalt (B1)/zinc blende (B3) phase causes a structural phase change to a cesium chloride (B2)/rocksalt (B1) phase. In order to understand them and other standing changes accompanying B1 to B2 transitions, it is necessary to study the thermodynamics, kinetics and structural aspects of the polymorphic phase transitions. The present section is devoted to such detailed description of these phenomena.

5.2.5 Pressure-Induced Phase Transitions

The pressure variation causes the overlap of the outer electron shells and this creates the transfer (or exchange) of charge which, in turn, changes the properties of the material. This results in structural phase transition and changes in the elastic behaviour of the materials. These observations are similar ones caused due to pressure variations have resulted a vital role for pressure in the study of properties of number of compounds, alloys etc. The present study is on the structural phase transition and their elastic behaviour due to the pressure on alkaline earth chalcogenides of RS structure.

The effect of high pressure influences many of the physical properties of solids. Bridgman studied the field of high pressure physics broadly and exposed many unique phenomena. In recent years, high pressure has made rapid advances both with respect to the limit of the pressure range available and to techniques. The most effective results of high pressure are the phase transition phenomena. High-pressure polymorphism appears to be a rule rather than an exception. The less efficiently packed a structure, the higher is the probability of a pressure-induced phase change. Even closed packed structures, such as face cantered cubic (FCC) and hexagonal closed packed (HCP), exhibit phase transition under pressure accompanied with small volume changes.

Pressure as a thermodynamic variable provides a means of changing the interatomic distances in materials in a controlled manner. Experimenters can now change the densities of condensed matter by upward of an entire order of magnitude, and thereby impart dramatic changes in physical and chemical properties of materials. The decrease in the volume brought about by pressure can be more than 50% compared with a change of only a few per cent resulting from a temperature increase in materials up to their melting point. Application of pressure has thus led to the discovery of phenomenon like the solid–solid critical point, molecular dissociation, new structural phases, etc. Pressure-induced phase transition in materials is more of a rule than an exception. A test on the utility of any classification is to see whether it enables us to understand a transition being studied.

During the past few decades, there has been a great deal of research in the area of pressure-induced phase transitions, in both the experimental techniques and the theoretical methods. In the static high-pressure experiments, the advances in the diamond anvil cell (DAC) technology and X-ray diffraction techniques with the use of synchrotron radiation sources have facilitated the detection of new phase transitions in solids under mega bar pressures. Also, in shock wave experiments it has become possible to detect phase transitions that are accompanied by small volume changes (0-5%) by using a new technique based on the observation of a discontinuity in the measured sound velocity as a function of the peak pressure in the shocked state.

On the theoretical side, there are various approaches to analyze the pressure-induced phase transitions. These vary from the use of the empirical methods based on the general trends, to the application of the first principle calculations. With the increase in the computational speed and improvement in the theory, the *ab initio* band structure methods have acquired the capability to calculate small structural energy differences between different phases, so that successful prediction of the structural phase transitions are presently possible. Molecular dynamics methods are also becoming increasingly popular for analyzing the mechanisms of phase transitions.

Under pressure, most of the observed phase transitions are thought to be diffusion-less in nature as pressure tends to suppress the large-scale diffusion of atoms in the solid state. This is because compression very sharply increases the repulsive forces between neighboring atoms, with the result that the height of the potential barrier, which must be overcome by the atoms in order to diffuse, increases. So the continued increase in pressure gradually distorts the structure, molecular shapes, and orientations and prepares the structure for undergoing phase transition with very little final displacement of the atoms. These small terminal displacements are usually reversible (even in crystalline to amorphous transitions) and maintain atom-by-atom correspondence between the positions in the parent and the product structures.

Because of this correspondence, the parent and the product phases display interesting symmetry relationships and the phase transitions can be analysed using Landau theory.

5.3 METHOD OF CALCULATIONS

The Gibbs free energy [GFE] at zero temperature and pressure P for BX (X=1,2) phase is:

$$\langle G_{BX}(r)\rangle = \frac{-\alpha_M{}^X Z^2 e^2}{r^X} - \frac{12\alpha_M{}^X Z e^2 \left(f_m(r)\right)}{r^X} - \left\{\frac{r^{X8}(C)^X + (D)^X (r)^{(X)6}}{r^{X8}(r)^{(X)6}}\right\} \tag{5.1}$$
$$+ 6b\beta_{ij}\exp\left[\left(r_i + r_j - r^X\right)/\rho\right] + 6b\beta_{ii}\exp\left[\left(2r_i - Y_X r^X\right)/\rho\right]$$
$$+ 6b\beta_{jj}\exp\left[\left(2r_j - Y_X r^X\right)/\rho\right] + PV_{BX}\left(r^X\right)$$

The symbols have their usual meaning.

5.4 RESULTS AND DISCUSSION

The structural stability is determined by calculating the GFE for both phases using the technique of minimization. The calculated PT for PuS is 72.5 GPa and 33.5 for PuSe is shown in Figure 5.1. Moreover, the compression in volume have been obtained

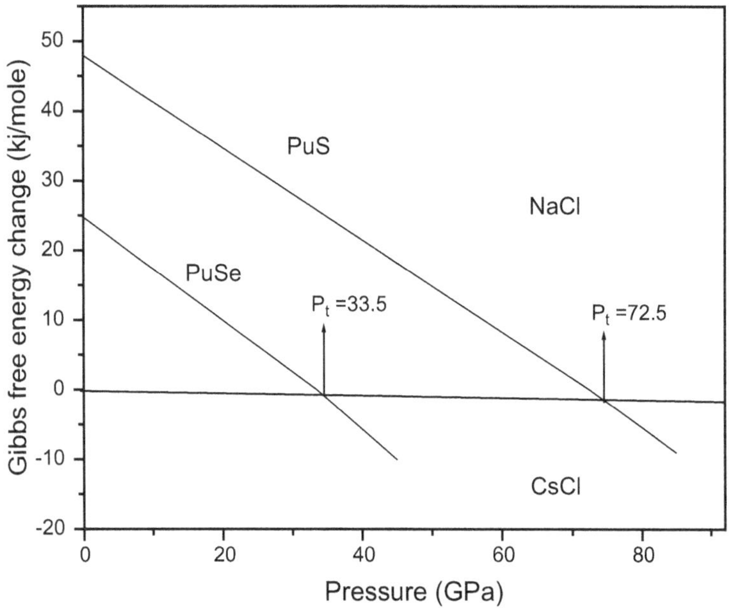

FIGURE 5.1 Variation of Gibbs free energy change with pressure.

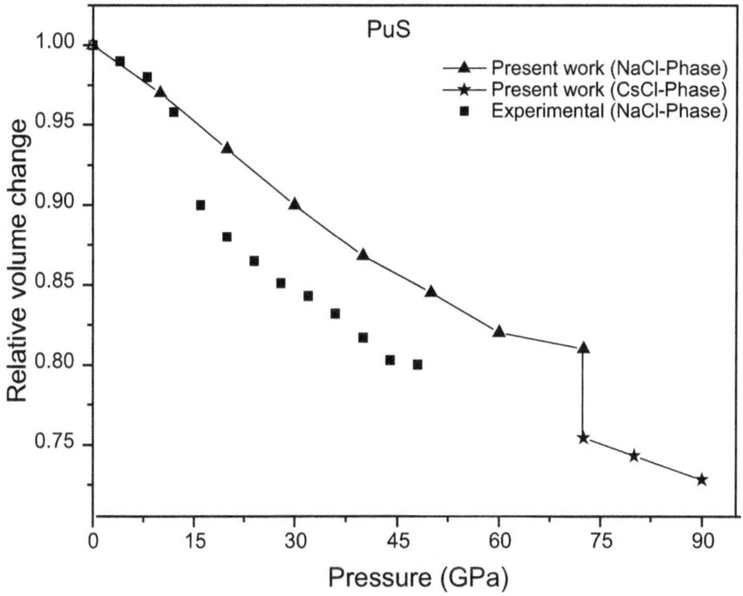

FIGURE 5.2 Variation of relative volume change with pressure for PuS.

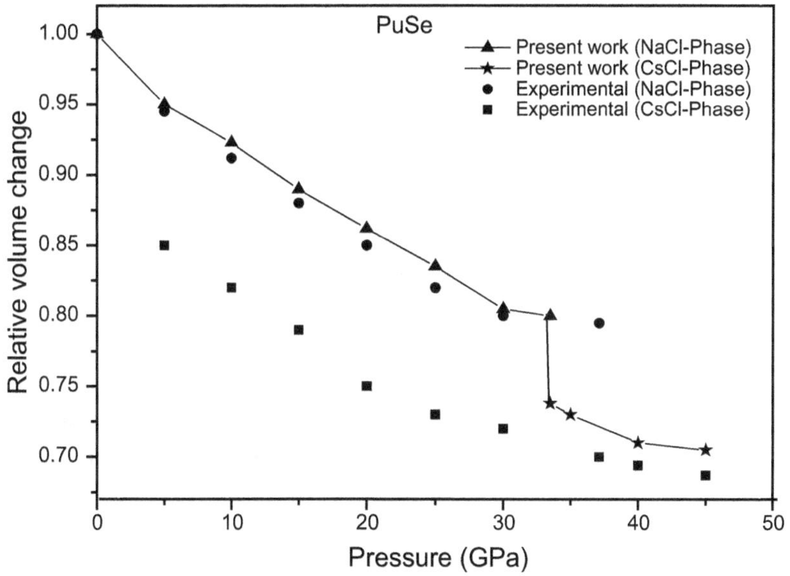

FIGURE 5.3 Variation of relative volume change with pressure.

with respect to various pressures. These results are presented in Figure 5.2 and 5.3 for PuS, PuSe. These results have reported and compared [10, 14] in Table 5.2. VC % of PuS and PuSe are 5.6 and 6.2 which are very close to experimental and theoretical

TABLE 5.2
Structural Properties

Compound	Phase transition Pressure (GPa)	Volume Collapse
PuS	72.5	5.6
Others	105.0[a]	3.6[a]
Expt.	>60.0[b]	–
PuSe	33.5	6.2
Others	37.1[a]	5.5[a]
Expt.	35.0[c]	11.0[c]

[a] ref [10]
[b] ref [14]
[c] ref [15]

TABLE 5.3
Elastic Constants and Bulk Modulus

Compound	C_{11} (GPa)	C_{12}(GPa)	C_{44}(GPa)	B(GPa)
PuS	245	55	63	118.3
Others	238[a]	53[a]	54[a]	115[a]
Expt.	–	–	–	119[b]
PuSe	218	42	56	100.6
Others	210[a]	34[a]	34[a]	92[a]
Expt.	–	–	–	98[c]

[a] ref [10]
[b] ref [15]
[c] ref [14]

results. Furthermore, the elastic properties [11–13] of present compounds are also taken into account. These results are presented in Table 5.3.

5.5 CONCLUSION

Finally, we have predicted the phase diagrams of PuS and PuSe, including the PT and VC. To optimize the structural properties, we have determined PT at which observed compounds undertaken transformation from $NaCl_{structure}$ to $CsCl_{structure}$. EC are also calculated successfully using this model.

REFERENCES

1. M. Mendik, P. Wachter, J.C. Spirlet, and J. Rebizant, *Physica B* 186–188 (1993) 678.
2. M.S.S. Brooks, *J. Magn. Magn. Mater.* 63&64 (1987) 649.
3. J.M. Fournier, E. Pleska, J. Chiapusio, J. Rossat-Mignod, J. Rebizant, J.C. Spirlet, and O. Vogt, *Physica B* 163 (1990) 493.
4. P. Wachter, F. Marabelli, and B. Bucher, *Phys. Rev. B* 43 (1991) 11136.

5. G.R. Stewart, R.G. Haire, J.C. Spirlet, and J. Rebizant, *J. Alloys Compd.* 177 (1991) 167.
6. P. M. Oppeneer, T. Kraft, and M. S. S. Brooks, *Phys. Rev. B* 61 (2016) 12824.
7. A. Hasegawa, and H. Yamagami, *J. Magn. Magn. Mater.* 65 (1992) 104.
8. M.S.S. Brooks, *J. Magn. Magn. Mater.* 63&64 (1987) 649.
9. A. Shick, L. Havela, T. Gouder, and J. Rebizant, *J. Nuclear Mater.* 385 (2009) 21.
10. V. Shrivastava, and S.P. Sanyal, *J. AlloysCompd.* 366 (2004) 15.
11. P. Bhardwaj, and S. Singh, *Eur. Phys. J. B* 86 (2013) 144.
12. P. Bhardwaj, and S. Singh, *J. Phase Transitions* 85 (9) (2012) 791.
13. P. Bhardwaj, and S. Singh, *Cent. Eur. J. Chem.* 10 (5) (2012) 1391.
14. U. Benedict, *J. Alloys Comp.* 223 (1995) 216.
15. T. LeBihan, S. Heathman, and J. Rebizant, *High Pressure Res.* 15 (1997) 387.
16. P. Bhardwaj, R.K. Bhardwaj, and A.P. Mishra, *ACTA Phys. Pol.* 137 (6) (2020) 1193–1195.
17. R.K. Bhardwaj, *Mater. Sci.-Pol.* 34 (1) (2016) 45–52.
18. R.K. Bhardwaj, P. Bhardwaj, and S. Narayan, *Electronic Study of Rare Earth Intermetallic Compound Materials Today Proceedings*, Vol. 29, P2, 633–637 (2020).
19. R.K. Bhardwaj, and C. K. Jaiswal, *Investigation of Structural Phase Transition of Ionic Compound Materials Today Proceedings*, Vol. 29, P2, 595–598 (2020).
20. R.K. Bhardwaj, P. Bhardwaj, and S. Singh, *Structural Properties of Rare Earth Chalcogenides, AIP Conf. Proc.* Vol. 1728, 020699-1-020699-4, 020699-1-4 (2016).
21. P. Bhardwaj, R.K. Bhardwaj, and Sadhna Singh, *Procedia Comput. Sci.* 57(2015) 57–64.

6 The Dynamics of a Continuous Innovation Diffusion Model with Advertisements as Well as Interpersonal Communications

Rakesh Kumar

CONTENTS

6.1 INTRODUCTION

Innovation diffusion is an unending activity that relies upon a network and the dispersion. It means the practice that transmits between the members of the social arrangement through a particular channel across time [1, 2]. The phenomenon of diffusion of innovation has been widely studied in the context of various disciplines, including technology and product marketing, medical epidemiology,

DOI: 10.1201/9781003089636-6

agriculture, ecology, sociology, and anthropology. Models that rely on diffusion theory to predict the adoption of an innovation are called diffusion models. An innovation is practice, object, or an idea which is adopted as new by an adopter or a group of adopters. Although under Roger's model of a new product, diffusion was widely accepted in the literature markets yet it had shortcomings. The shortcomings of Roger's model were first explored by Bass in the Bass model (1969) [3], in which he presented a new model of innovation diffusion theory as compared to Roger's model. Over the past six decades, many researchers have a significant focus on the aspects of individual adoption of the latest innovative products, including its diffusion into a social network. The Bass model has been in use to predict the diffusion of innovation in a population by emphasizing the rate of communication between adopters and non-adopters as an internal process, The Bass model has been increasingly used to understand the pattern of growth of adopters by the influence of mass media, including print and electronic coverage or various external factors, word-of-mouth interactions and other kind of influences [4–10]. The model incorporating all the above-mentioned factors developed by Bass is as follows:

$$\frac{dA(t)}{dt} = \left(p + q\frac{A(t)}{m}\right)(m - A(t)),$$

where $A(t)$ is the cumulative number of adopter at time t, m is the total population of potential adopters, p is the coefficient of innovation and q is the coefficient of imitation. The first term in the equation denotes the adoption by innovators and the second term denotes the adoption by imitators.

The informative stage and a final adoption stage have also been incorporated into the Bass model in [11]. The Bass approach is a development of the other two factors and assumes that potential non-users (or adoption units) are getting affected in their purchase behaviour by information sources of mass-media communications and word of mouth. In the Bass model, it has been assumed that the number of potential adopters (i.e. the total population) remains unchanged in the diffusion process, yet this does not match the real-world scenario, where the changes in the demography of the population may be caused by immigration, emigration as well as due to its intrinsic growth (i.e. due to birth and death processes in a population) [12–13].

The unit of analysis for Cellular Automata is represented by interpersonal links (edges): a standard edge between two different agents or a reflexive edge if we refer to a single agent, taking into account "auto-communication" (Guseo and Guidolin [14]). Mahajan and Sharma [15] accounting the shortcomings of the Bass proposal for parameter estimation, suggested a simple algebraic estimation procedure. This method is applicable not only for the Bass model parameter estimation but also for the other diffusion models. This method does not employ period by period time-series data and it may not provide the best fit to the data. On the other hand, the method can be helpful when data up to the point of inflection are missing and when absence or limited data are given. An alternative method used for the parameter estimation of diffusion models is the Augmented Kalman Filter with Continuous State and Discrete Observations [16]. This procedure claims better applicability to the models and is independent of the constraints of either the model's structure or the parameter's nature. More specifically, it can be applied either in time-invariant and time-variant models and it can also be used in cases of both solvable and unsolvable diffusion models. AKF(C-D) approach can be used for estimating parameters that are deterministic or stochastic and observation errors can be incorporated [16]. The algorithm of AKF(C-D) is a Bayesian-updating procedure. A comparison of this method in relation to the other methods showed that in most of the cases it outperforms them.

With regard to the analysis, the Bass model with maximum likelihood estimation (MLE) seemed to be correct for the dishwashers. For clothes dryers, air conditioners and color TVs, MLE provided a better fit to the data and the one-step-ahead sales forecasts were predicted with validity. The data for the consumer products were survey data. In the cases of the medical products, the MLE possessed correct signs and conceivable values, concluding a good fit to the adoption data. The data for the medical equipment were sample data from hospitals. A logical consequence of the observations showed that more data, acquired by lengthening the data collection period or collecting more data for the same, could increase the reliability of the parameters estimations and creation of confidence regions. In all cases, the individual adoption times are not known making the use of a histogram with the number of individuals falling in each time interval necessary to fit the data. Srinivasan and Mason [17] apply a Bass model of innovation diffusion and focus on the fit and validity of a parameter estimation method; Nonlinear Least Squares (NLS). This can be characterized as a logical reaction to the Schmittlein and Mahajan paper. The reason is that Srinivasan and Mason [17] apply the same model on the same products with the Schmittlein and Mahajan [18] paper, in order to combine the Maximum Likelihood Estimation (MLE) method to NLS and prove NLS's effectiveness. Focusing on the communication channels by Roger's (1983) [2], the paper in [18] is interesting as it includes in its analysis a mass media communication channel itself (color TVs). This excludes the mass-media communication channels as a mean to transfer information for products. Considering the time period of the diffusion of all the consumer products, it is more probable that the diffusion of the innovations occurred via interpersonal communication channels. Since the only information is based on the Statistical Abstracts of the United States, and the records demonstrate the diffusion of the products all over the USA, a distinction of the communication channels according to their sources is risky. Probable among the medical products diffused in hospital, some might suggest that the communication channels are localite, because the products are very specialized and the hospitals are a closed market. For the house consumer products, an argument to support the diffusion of the products via localite communication channels is based on the habit of consumers to 'advertise' in their local community the products they purchase. We should not forget that at that time there were no mass media expanded and broadly used in the households and organizations as they are nowadays.

Kennedy [19] focused on the problematic aspect of the 'managerial intuition', which is not correctly estimated usually in the model. The term managerial intuition refers to extraneous various sources that specify the parameter values. The aim is to 'extend the mixed estimation technique to handle stochastic prior information' as an acknowledgement that the validity of the value of the managerial intuition is uncertain. There are mostly some not so reliable methods applied to calculate the managerial intuition (term m in the model that defines the total number of first purchasers). These are the Bayesian method, or the use of a common value for similar products (i.e. consumer products) when the value seems to work for the most cases applied. The aim of this chapter is to show that the technique proposed is a good and simple-to-compute alternative to the Bayesian method. Mathematically speaking, the equation used is the typical Bass model. The additive value of the chapter is the presentation of their proposed parameter estimation method.

In an epidemic model as the contamination spread from an association of the infected individual to premature individual is investigated in [20]. Thus, in that way, in the innovation diffusion process, the non-adopted person can become the user of innovation by the direct interaction with the adopted person [21]. The age-structured mathematical modelling approach to short-term diffusion of electronic commerce in Spain was studied in further the short-term trends of e-commerce [22]. In the communication market, the researchers have investigated the dynamics of competitive multi-innovations and its application [23–25].

From the literature survey, it can be easily seen that remarkable work has been completed in the direction of the product expansion. However, more issues are to be resolved yet. Here, the author will focus on the most important aspect in the potential market, i.e., the role of financial conditions of the population in the innovation diffusion process of high-value products. In this world, every individual has its own spending capacity. Few of them can purchase the new products irrespective of their prices; on the other side, however, some of the individuals are needed to think during the diffusion process, and they cannot buy the product in a split second. It is expected that they can buy the product, later on, when they remain financially stable. It means adoption of the high-end innovations will totally depend on the financial positions of the people. This is not practically possible in the society that the costlier things can be bought by the poor population. The distinction between the low-income and high-income groups was missed in the previous works, i.e., a wide research gap exists. This is an important aspect in the present market scenario for the diffusion of a high-valued product in the market. Keeping in mind these scenarios, a delay differential equations model is proposed and further analysed by using the classical methods of innovation diffusion modelling, and the model includes the effect of information received through mutual word-of-mouth interactions between the population of non-adopters and adopters, and by the external advertisements (print and electronic), which leave the non-adopter population virtually prompt to adopt the high-valued innovation in the markets. Hence, the moving of the non-adopter population to the adopter class will occur by the effect of interactions and advertisements after a certain time lag. The effects of other demographic processes of the population such as emigration rate, death rate, and so on have also been included in the innovation diffusion.

This chapter is managed as follows: In Section 6.2, a nonlinear innovation diffusion model incorporating time delay as control strategy, external and internal influences are presented. In Section 6.3, the basic preliminaries, including the positivity and boundedness of the system, are checked. The next section related to discuss the concept of existence of equilibrium points and basic influence numbers. Section 6.5 then deals with the dynamical behaviour of the time delayed model, and helped to find the local asymptotic stability conditions of adopter-free and interior equilibrium points. The model is investigated for the Hopf bifurcation by using bifurcation theory in the next section. Section 6.7 deals with the sensitivity analysis at the positive steady state. In Section 6.8, the numerical simulations are executed in the support of analytical outcomes. Finally, in the concluding section, the basic results of mathematical findings are presented with their significance to real life scenarios.

6.2 MATHEMATICAL MODEL

The financial status of the people may vary with the passage of time. The diffusion model system is constituted by considering the costlier products. The products, which are not in the range of everyone in the society, is taken into consideration. The total population is sub-divided into three categories, namely, the non-adopters population of the low-income group with density $N_1(t)$, the non-adopters of the high-income group with density $N_2(t)$, and the adopter population with density $A(t)$, where t denotes the time. In order to make the model effective and increase its similarity to real-life situations, the demographic processes of a population have been considered into the classical Bass model. The assumptions for the formulation of the mathematical model are detailed below:

1. Suppose that Λ_1 and Λ_2 are the new recruitment rates of non-adopter populations $N_1(t)$ and $N_2(t)$ respectively.
2. The financial status of the people may vary with the passage of time. It means that the non-adopter population of the low-income group $N_1(t)$ can be the population of the high-income group $N_2(t)$ in future times and vice versa. This means that their populations can migrate between these groups according to their current financial positions. Assume that ξ_1 is the rate at which the non-adopter population of the low-income group $N_1(t)$ join the population of the high-income group $N_2(t)$ and ξ_2 is the rate at which the non-adopter population of the high-income group $N_2(t)$ join the population of the low-income group $N_1(t)$.
3. As we are taking the high-valued innovations, so it is supposed that only the non-adopters population of the high-income group $N_2(t)$ is capable enough to adopt the product.
4. Suppose that the variable external influences (advertisements) and variable internal influences (word of mouth) directly affect the decision about the adoption. Therefore, the rate of change of the adopters depends on both external and internal influences.
5. Let α be the rate of interactions (word of mouth) occurred between the non-adopter population $N_2(t)$ and the adopter population $A(t)$. Also let β be parameter which has been used for cumulative density of variable external influences (advertisements), to affect the decisions of the non-adopters population $N_2(t)$.

6. The shifting of non-adopters to the adopter population is not instantaneous, but takes a small time, defined by the average evaluation period τ. Moreover, an individual who has information of the product at time $t - \tau$ may leave the evaluation class in the interval $[t - \tau, t]$ due to loss of interest or his death at the present time t. It is also supposed that the individuals who pass the evaluation stage will definitely move to the adopter population class, so $(\alpha A(t - \tau) + \beta)N(t - \tau)$ is the non-adopters directly transferring to the adopter population having information about the product at time $t - \tau$ by internal and external influences, who join the adopter class $A(t)$ at time t.
7. Let v be the rate at which the adopters joining back to the non-adopter population class $N_2(t)$ because of becoming dissatisfied by the performance of the product that, of course, may come again to join later depending upon all the circumstances. Suppose that δ is the death and emigration rate of the various populations $N_1(t)$, $N_2(t)$ and $A(t)$.

Here, in the following system (6.1), to observe the impact of τ in understanding the patterns of the continuous dynamics of the non-adopter and adopter populations, the author consider it as a control parameter. Thus, the mathematical form of the model is given as below:

$$\begin{aligned}
\frac{dN_1}{dt} &= \Lambda_1 - \xi_1 N_1(t) + \xi_2 N_2(t) - \delta N_1(t),\\
\frac{dN_2}{dt} &= \Lambda_2 + \xi_1 N_1(t) - \xi_2 N_2(t) - \left(\alpha A(t-\tau) + \beta\right) N_2(t-\tau) + vA(t) - \delta N_2(t),\\
\frac{dA}{dt} &= \left(\alpha A(t-\tau) + \beta\right) N_2(t-\tau) - (\delta + v) A(t),
\end{aligned} \tag{6.1}$$

These equations are to be solved subject to the initial conditions

$$N_1(\theta) = \phi_1(\theta),\ N_2(\theta) = \phi_2(\theta),\ A(\theta) = \phi_3(\theta);\ \phi_i(0) > 0,\ \forall i = 1, 2, 3 \tag{6.2}$$

The functions $\phi_1(\theta)$, $\phi_2(\theta)$, $\phi_3(\theta)$ are continuous functions which are bounded in the interval$[-\tau, 0]$. Specifically, $\phi_1(\theta)$, $\phi_2(\theta)$, $\phi_3(\theta) \in C([-\tau, 0], R_+^3)$, the Banach space of functions mapping the interval $[-\tau, 0]$ into $\mathfrak{R}_+^3 = \{(y_1, y_2, y_3): y_i > 0, i = 1, 2, 3\}$, which are continuous. Applying the basic results of FDE [26], make a note that all the solutions $N_1(\theta)$, $N_2(\theta)$, $A(\theta)$ of the IVP are always unique and nonnegative on $[0, +\infty)$.

6.3 BASIC PRELIMINARIES

For a well-posed mathematical problem, there should be positive solutions of the model (6.1) lying in the boundary of a region. For this, the author presented the subsequent lemmas for proving the boundedness and non-negativity of the proposed model (6.1).

Lemma 6.1: All solutions of the model (6.1) subject to the initial values (6.2) are non-negative, ∀t ≥ 0.

Proof. For $t \in [0, \tau]$, the first equation of model (6.1) can be re-written as

$$\frac{dN_1}{dt} \geq_1 -(\xi_1 + \delta)N_1(t).$$

This gives us

$$N_1(t) \geq N_1(0)\exp\left[-\int_0^t (\xi_1 + \delta)\,dv\right] > 0.$$

The second equation of the system can also be written as

$$\frac{dN_2}{dt} \geq -(\xi_2 + \delta)N_2(t) - (\alpha A(t-\tau) + \beta)N_2(t-\tau)$$
$$\frac{dN_2}{dt} + (\xi_2 + \delta)N_2(t) = -(\alpha A(t-\tau) + \beta)N_2(t-\tau)$$

On solving, we shall have

$$N_2(t) \geq e^{-\int_0^t (\xi_2+\delta)d\theta}\left[N_2(0) + \int_0^t \left\{(\alpha A(\theta-\tau) + \beta)N_2(\theta-\tau)\right\}e^{\int_0^\theta (\xi_2+\delta)ds}\,d\theta\right] > 0,$$

$$\forall t \in [0, \tau].$$

The third equation of the model (6.1) gives

$$\frac{dA}{dt} + (\delta + v)A(t) \geq (\alpha A(t-\tau) + \beta)N_2(t-\tau),$$

Solving the equation, we shall have

$$A(t) \geq e^{-\int_0^t (\delta+v)d\theta}\left[A(0) + \int_0^t \left\{(\alpha A(\theta-\tau) + \beta)N_2(\theta-\tau)\right\}e^{\int_0^\theta (\delta+v)ds}\,d\theta\right] > 0, \text{ for } t \in [0, \tau].$$

Similarly, one can take the successive intervals $[\tau, 2\tau]$, $[2\tau, 3\tau]$, ……$[n\tau, (n+1)\tau]$; $n \in \mathbb{N}$ [27], and make a result that all the solutions of the model (6.1) always positive $\forall t \geq 0$.

Lemma 6.2: The non-negative solutions of model (6.1) with reference to the mentioned initial values (6.2) in R_+^3 are bounded.

Proof. Suppose that $\Omega(t) = N_1(t) + N_2(t) + A(t)$, $\forall\, t \geq 0$.
Differentiating w.r.t. t, and obtain

$$\frac{d\Omega}{dt} = \Lambda_1 + \Lambda_2 - \delta\left(N_1(t) + N_2(t) + A(t)\right),$$

$$\frac{d\Omega}{dt} = \Lambda_1 + \Lambda_2 - \delta\Omega,$$

$$\text{Therefore}\ \frac{d\Omega}{dt} + \delta\Omega = \Lambda_1 + \Lambda_2, \text{for any}\ \delta > 0.$$

Using an important result from [28] on differential inequalities, one can obtain

$$0 \leq \Omega(t) \leq \frac{\Lambda_1 + \Lambda_2}{\delta}\left(1 - e^{-\delta t}\right) + \Omega(0)e^{-\delta t}, \text{for any}\ \delta > 0.$$

$$\text{As a limiting form}\ t \to \infty, 0 \leq \Omega(N_1, N_2, A) \leq \frac{\Lambda_1 + \Lambda_2}{\delta}.$$

Thus, all solutions of the model (6.1) which start in R_+^3 will always be in the region

$$\Sigma = \left\{\left(N_1(t), N_2(t), A(t)\right) : 0 \leq \Omega(t) \leq \frac{\Lambda_1 + \Lambda_2}{\delta} + \epsilon, \text{ for any } \epsilon > 0\right\}.$$

Thus $N_1(t)$, $N_2(t)$, $A(t)$ will always have upper bound in the region Σ, $\forall\, t \geq 0$.

6.4 EXISTENCE OF EQUILIBRIA AND BASIC INFLUENCE NUMBER

The model (6.1) must possess the below mentioned two equilibrium points:

(i) $E^0 = (N_1(t), N_2(t), 0)$, which is an adopter free equilibrium point, and where

$$N_1(t) = \frac{2\Lambda_1\xi_2 + \Lambda_1(\beta + \delta)}{\delta(\xi_2 + \beta + \delta) + \xi_1(\beta + \delta)}, N_2(t) = \frac{2\Lambda_1\xi_1 + \Lambda_1\delta}{\delta(\xi_2 + \beta + \delta) + \xi_1(\beta + \delta)}.$$

(ii) $E^* = (N_1^*(t), N_2^*(t), A^*(t))$ is the non-negative equilibrium.

The existence of adopter free point E^0 will be trivial. So, the detailed discussion about the possibility of non-negative point E^* must be there. Before the presence of E^*, the author shall firstly find out the formula of basic influence number. Basic influence number is the remarkable edge parameter, which outlines the mathematical model related to the adoption of the new product. Basic influence number is denoted by R_A and recognized as the part of consequent adoption constructed by the earlier adopter, and it is correlated with basic reproduction number from epidemiology [29].

The advantage of this matrix is to state either the innovative products will spread to population or not. Assume that f and v are vectors to signify the new users by the personal impact of the user population on the non-user population and enduring transfer terms, sequentially from model equations of user sections [29]. For our model the matrices f and v are given as:

$$f = \begin{bmatrix} \alpha N_2 A \\ 0 \\ 0 \end{bmatrix}, v = \begin{bmatrix} (\delta + v)A - \beta N_2 \\ -\Lambda_2 - \xi_1 N_1(t) + \xi_2 N_2(t) + (\alpha A(t-\tau) + \beta) N_2(t-\tau) - vA(t) + \delta N_2(t) \\ -\Lambda_1 + \xi_1 N_1(t) - \xi_2 N_2(t) + \delta N_1(t) \end{bmatrix}.$$

Assume F is the Jacobian of matrix f at adopter free steady state E^0

$$F = [\alpha N_2].$$

Also, V is taking the following form, which is the Jacobian of v,

$$V = [(\delta + v)].$$

Now, the spectral radius of (FV^{-1}) is the basic influence number R_A for the model given by $R_A = \dfrac{\alpha \Lambda_1 (2\xi_1 + \delta)}{(\delta + v)[\delta(\xi_2 + \beta + \delta) + \xi_1(\beta + \delta)]}$, which is the largest eigenvalue of (FV^{-1}).

Further, the solution of the subsequent set of mathematical equations provides the interior equilibrium point $E^* = (N_1^*(t), N_2^*(t), A^*(t))$ of the proposed model system (6.1):

$$\begin{cases} \Lambda_1 - \xi_1 N_1(t) + \xi_2 N_2(t) - \delta N_1(t) = 0, \\ \Lambda_2 + \xi_1 N_1(t) - \xi_2 N_2(t) - (\alpha A(t) + \beta) N_2(t) + vA(t) - \delta N_2(t) = 0, \\ (\alpha A(t) + \beta) N_2(t) - (\delta + v) A(t) = 0. \end{cases} \tag{6.3}$$

From Equation (6.3), we obtain

$$N_2^* = \frac{(\delta + v)A}{(\alpha A + \beta)}.$$

Similarly, from (6.3), and using the value of N_2^*, we can have $N_1^* = \dfrac{\Lambda_1(\alpha A + \beta) + (\delta + v)\xi_2 A}{(\alpha A + \beta)(\xi_1 + \delta)}$, and using these values in (6.3), one can get a polynomial equation in A^* of order two as below:

$$F(A^*) = X_1 A^{*2} + X_2 A^* + X_3 = 0, \tag{6.4}$$

$$\text{where}\begin{cases} X_1 = -\alpha\delta\left(\xi_1+\delta\right), \\ X_2 = \alpha\Lambda_1\xi_1 + \left(\alpha\Lambda_2 - \beta\delta - \delta\left(\delta+v\right)\right)\left(\xi_1+\delta\right) - \xi_2\delta\left(\delta+v\right), \\ X_3 = \beta\Lambda_2\left(\xi_1+\delta\right) + \beta\Lambda_1\xi_1. \end{cases}$$

It is clear from the values of X_1, X_2, and X_3 that X_1 will always remain negative and X_3 is always positive. Thus, by Descartes's rule, Equation (6.4) will always have a unique non-negative root.

6.5 STABILITY ANALYSIS OF VARIOUS EQUILIBRIA

This section constructed with the stability analysis of two equilibria, namely adopter free and non-negative of the model system (6.1).

6.5.1 Stability of Adopter Free Equilibrium Point E^0

The local stability of the adopter free equilibrium point $E^0 = (N_1(t), N_2(t), 0)$ is detailed in the forthcoming lemma:

Lemma 6.3: The point $E^0 = (N_1(t), N_2(t), 0)$ is LAS when $R_A < 1$ and remains unstable for $R_A > 1$.

Proof. The variational matrix of the model (6.1) is given by

$$J^0 = \begin{bmatrix} -\xi_1 - \delta - \lambda & \xi_2 & 0 \\ \xi_1 & -\xi_2 - \beta e^{-\lambda\tau} - \delta - \lambda & -\alpha N_2 e^{-\lambda\tau} + v \\ 0 & \beta e^{-\lambda\tau} & \alpha N_2 e^{-\lambda\tau} - \left(\delta+v\right) - \lambda \end{bmatrix}$$

The characteristic equation obtained from J^0 for the steady-state

$$E^0 = \left(N_1\left(t\right), N_2\left(t\right), 0\right) \text{ is as follows :}$$

$$\left(\lambda+\xi_1+\delta\right)\left(\lambda+\xi_2+\beta e^{-\lambda\tau}+\delta\right)\left(\lambda-\alpha N_2 e^{-\lambda\tau}+\delta+v\right)=0.$$

From this characteristic equation, we get

$$\lambda = -\xi_1 - \delta, -\xi_2 - \beta e^{-\lambda\tau} - \delta, \alpha N_2 e^{-\lambda\tau} - \delta - v.$$

For the local asymptotic stability, all the eigenvalues of λ should have negative real parts. So, from the above outputs, the first two values of λ are negative and third eigenvalue is conditionally negative. The third value is negative if $\alpha N_2 e^{-}\lambda\tau - \delta - v < 0$, i.e., if

$$R_A = \frac{\alpha\Lambda_1\left(2\xi_1+\delta\right)}{\left(\delta+v\right)\left[\delta\left(\xi_2+\beta+\delta\right)+\xi_1\left(\beta+\delta\right)\right]} < 1.$$

Thus, it has been proved that adopter-free equilibrium E^0 is LAS if the basic influence number R_A is less than one.

6.5.2 Stability of Interior Equilibrium Point E^*

For the delayed innovation diffusion model (6.1), the variational matrix obtained about the interior point E^* is given as below:

$$J^* = \begin{bmatrix} -\xi_1 - \delta - \lambda & \xi_2 & 0 \\ \xi_1 & -\xi_2 - (\alpha A^* + \beta)e^{-\lambda\tau} - \delta - \lambda & -\alpha N_2^* e^{-\lambda\tau} + v \\ 0 & (\alpha A^* + \beta)e^{-\lambda\tau} & \alpha N_2^* e^{-\lambda\tau} - (\delta + v) - \lambda \end{bmatrix}$$

The characteristic equation for J^* at E^* is obtained by taking the determinant

$$|J^* - \lambda I| = 0 \text{ as,}$$

$$(\lambda^3 + P_1\lambda^2 + P_2\lambda + P_3) + (Q_1\lambda^2 + Q_2\lambda + Q_3)e^{-\lambda\tau} = 0, \tag{6.5}$$

where

$$\begin{cases} P_1 = \xi_1 + \xi_2 + 3\delta + v; \\ P_2 = (\xi_1 + \delta)(\delta + v) + \xi_1(2\delta + v) + \delta(\xi_2 + 2\delta + v); \\ P_3 = \xi_1\delta(\delta + v) + \delta(\xi_2 + \delta)(\delta + v); \\ Q_1 = \alpha(A^* - N_2^*) + \beta; \\ Q_2 = \left[\alpha(A^* - N_2^*) + \beta\right](\xi_1 + \delta) + \delta(\alpha A^* + \beta) - \alpha N_2^*(\xi_2 + \delta); \\ Q_3 = \left[\delta(\alpha A^* + \beta) - \alpha N_2^*(\xi_2 + \delta)\right](\xi_1 + \delta) + \alpha N_2^* \xi_1 \xi_2; \end{cases}$$

Let us check for the stability analysis of steady state E^* of the model (6.1).

Firstly, analyze the model in the absence of τ. Without parameter τ, Equation (6.5) becomes

$$\lambda^3 + (P_1 + Q_1)\lambda^2 + (P_2 + Q_2)\lambda + (P_3 + Q_3) = 0. \tag{6.6}$$

With the help of the Routh-Hurwitz criterion, all values of Equation (6.6) will have negative real parts, i.e., the positive steady state E^* is LAS provided (H_1) hold good, where

$$\begin{aligned} &H_1 : (i)\, P_1 + Q_1 > 0, (ii)\, P_2 + Q_2 > 0, (iii)\, P_3 + Q_3 > 0, \\ &\text{and (iv)} \quad (P_1 + Q_1)(P_2 + Q_2) - (P_3 + Q_3) > 0. \end{aligned} \tag{6.7}$$

6.6 HOPF-BIFURCATION ANALYSIS

In this section, the author proceed to investigate the stability of model (6.1) for $\tau > 0$, and the Hopf bifurcation analysis of the system about $E^* = (N_1^*(t), N_2^*(t), A^*(t))$ and it will be validated in the numerical simulation section.

Lemma 6.4: [21]

(i) The point E^* of (6.1) is absolutely stable iff the point E^* of the corresponding ODE model is asymptotically stable and the characteristic Equation (6.5) will have real roots for $\tau > 0$.
(ii) The equilibrium E^* of (1) is conditionally stable iff all roots of (6.5) have negative real parts for zero τ and $\exists\ \tau > 0$ so that the Equation (6.5) has a complex conjugate pair of purely imaginary values of the form $\pm i\omega$.

The following theorem may also be stated:

Theorem 6.5: The interior point E^* is conditionally stable if (H_1) hold good for the model (6.1).

To check, how the parameter τ (time delay) affect the conditions of the stability of positive steady state E^*, let τ be the bifurcation parameter. To find the instability incurred in the system (6.1) by a delay parameter τ, let $\lambda = i\omega$, for any $\omega > 0$, be a root of (6.5) $(\tau > 0)$. Put $\lambda = i\omega$ into Equation (6.5), and find out real part and imaginary part of the characteristic Equation (6.5),

$$RP: Q_2\omega\sin(\omega\tau) + (Q_3 - Q_1\omega^2)\cos(\omega\tau) = P_1\omega^2 - P_3, \tag{6.8.}$$

$$IP: Q_2\omega\cos(\omega\tau) + (Q_3 - Q_1\omega^2)\sin(\omega\tau) = \omega^3 - P_2\omega, \tag{6.9}$$

Solving Equations (6.8.) and (6.9) for sine and cosine terms, and using the formula of trigonometry $\sin^2\theta + \cos^2\theta = 1$, one can easily get after simple computations

$$\Phi(\rho) = \rho^3 + \Delta_1\rho^2 + \Delta_2\rho + \Delta_3 = 0, \tag{6.10}$$

where

$$\rho = \omega^2, \text{and } \Delta_1 = P_1^2 - 2P_2 - Q_1^2, \Delta_2 = P_2^2 - 2P_1P_3 + 2Q_1Q_3 - Q_2^2, \Delta_3 = P_3^2 - Q_3^2. \tag{6.11}$$

If there exists a minimum of one non-negative root of equation, then the characteristic Equation (6.5) will have purely imaginary roots. So, $\omega = \pm\sqrt{\rho_1}$ will be the pair of eigenvalues, which are complex conjugate, and this is possible if ρ_1 (say) is a positive real root of (6.10). Thus, the Hopf bifurcation might occur in the system (6.1).

The following lemma explains the possibility of the existence of a positive real root of (6.10):

Lemma 6.6: The equation $\Phi(\rho) = 0$ has a minimum one positive real root iff one of the given conditions are justified:

(i) $\Delta_3 < 0$, (ii) $\Delta_3 \geq 0$, $\Delta_1^2 - 3\Delta_2 > 0$, and $\rho_c > 0$ of $\Phi(\rho)$ exists with $\Phi(\rho_c) \leq 0$.

Thus, the model system (1) has purely imaginary eigenvalues if and only if conditions (i) or

(ii) in Lemma 6.6 are verified. Also Equation (6.10) can have at most three positive real roots, $\rho_i > 0$; $i = 1, 2, 3$, it is also possible that there may exist three purely imaginary pairs of eigenvalues, $\lambda_i = i\omega_i = \pm i\sqrt{\rho_i}, i = 1, 2, 3$. Let us now find the values of the time delay parameter τ_i corresponding to values of ω_i by applying ω_i into Equation (6.8.) and Equation (6.9) for the real and imaginary parts of the characteristic Equation (6.5). Solving for $\sin(\omega_i\tau)$ and $\cos(\omega_i\tau)$, one can easily obtain the following:

$$\sin(\omega_i\tau) = \frac{\left(P_1\omega_i^2 - P_3\right)Q_2\omega_i - \left(\omega_i^3 - P_2\omega_i\right)\left(Q_3 - Q_1\omega_i^2\right)}{Q_2^2\omega_i^2 - \left(Q_3 - Q_1\omega_i^2\right)^2} \tag{6.12}$$

$$\cos(\omega_i\tau) = \frac{\left(\omega_i^3 - P_2\omega_i\right)Q_2\omega_i - \left(P_1\omega_i^2 - P_3\right)\left(Q_3 - Q_1\omega_i^2\right)}{Q_2^2\omega_i^2 - \left(Q_3 - Q_1\omega_i^2\right)^2} \tag{6.13}$$

The threshold value of τ at which the system exhibits stability switch, i.e., values of (6.5) at which complex roots occur, can be obtained by using Equations (6.12) and (6.13) and get that

$$\tau_i^{(j)} = \frac{1}{\omega_i}\arctan\left[\frac{\left(P_1\omega_i^2 - P_3\right)Q_2\omega_i - \left(\omega_i^3 - P_2\omega_i\right)\left(Q_3 - Q_1\omega_i^2\right)}{\left(\omega_i^3 - P_2\omega_i\right)Q_2\omega_i - \left(P_1\omega_i^2 - P_3\right)\left(Q_3 - Q_1\omega_i^2\right)}\right] + \frac{2j\pi}{\omega_i}, \tag{6.14}$$

where $i = 0, 1, 2,$ and $j = 0, 1, 2, 3\ldots$ The least value of τ_0 at which the purely imaginary eigenvalues of the form $\lambda_0 = \pm\, i\omega_0$ occur is therefore given as below:

$$\tau_0 = \min_{0\le i\le 2,\, j\ge 0} \tau_i^{(j)},\ \tau_i^{(j)} > 0. \tag{6.15}$$

Now, let us find the condition of Hopf bifurcation for the system (6.1) around the interior point E^*. Taking τ as a bifurcation parameter and suppose that $\tau > 0$, $\lambda = \mu + i\omega$ is a root of Equation (6.5), where $\omega > 0$ is a real. Putting $\lambda = \mu + i\omega$ into (6.5), and separating R.P. and I.P., and obtained as follows:

$$\begin{aligned}&\mu^3 - 3\mu\omega^2 + P_1\left(\mu^2 - \omega^2\right) + P_2\mu + P_3 + \\ &\left[\left\{Q_1\left(\mu^2 - \omega^2\right) + Q_2\mu + Q_3\right\}\cos(\omega\tau) + \left(2Q_1\mu\omega + Q_2\omega\right)\sin(\omega\tau)\right]e^{-\mu\tau} = 0,\end{aligned} \tag{6.16}$$

and

$$\begin{aligned}&\left(-\omega^3 + 3\mu^2\omega + 2P_1\mu\omega + P_2\omega\right) + \\ &\left[-\left\{Q_1\left(\mu^2 - \omega^2\right) + Q_2\mu + Q_3\right\}\sin(\omega\tau) + \left(2Q_1\mu\omega + Q_2\omega\right)\cos(\omega\tau)\right]e^{-\mu\tau} = 0,\end{aligned} \tag{6.17}$$

The conditions of bifurcation have been used in the theory of Hopf bifurcations [33]. One has been proved in lemma (6.5). Differentiating Equations (6.16) and (6.17) w.r.t. τ and letting $\tau=\hat{\tau}, \omega=\hat{\omega}$, and $\mu=0$, the obtained expressions are as below:

$$E_1\left[\frac{d\mu}{d\tau}\right]_{\tau=\hat{\tau}} - E_2\left[\frac{d\omega}{d\tau}\right]_{\tau=\hat{\tau}} = G_1, \tag{6.18}$$

$$E_2\left[\frac{d\mu}{d\tau}\right]_{\tau=\hat{\tau}} + E_1\left[\frac{d\omega}{d\tau}\right]_{\tau=\hat{\tau}} = G_2, \tag{6.19}$$

Where

$$E_1=-3\hat{\omega}^2+P_2+\left\{Q_2+\tau\left(Q_1\omega^2-Q_3\right)\right\}\cos(\omega\tau)+\omega\left(2Q_1-Q_2\tau\right)\sin(\omega\tau),$$
$$E_2=2P_1\omega-\left\{Q_2+\tau\left(Q_1\omega^2-Q_3\right)\right\}\sin(\omega\tau)++\omega\left(2Q_1-Q_2\tau\right)\cos(\omega\tau),$$
$$G_1=\omega\left(Q_3-Q_1\omega^2\right)\sin(\omega\tau)-Q_2\omega^2\cos(\omega\tau),$$
$$G_2=\omega\left(Q_3-Q_1\omega^2\right)\cos(\omega\tau)+Q_2\omega^2\sin(\omega\tau).$$

Simplify Equations (6.18) and (6.19), and at $\tau=\hat{\tau}=\tau_0, \omega=\hat{\omega}^2=\omega_0{}^2$,

$$\left[\frac{d\mu}{d\tau}\right]_{\tau=\tau_0}=\frac{\hat{\omega}^2}{E_1{}^2+E_2{}^2}\left[\frac{d\Phi}{d\rho}\right]_{\omega=\omega_0{}^2}\neq 0,$$

Where $\Phi(\rho)$ is stated in Equation (6.10). Hence, the roots of the characteristic Equation (6.5) cross the vertical axis as the bifurcation parameter τ crosses over the threshold value. Hence, at the threshold value $\tau=\tau_0$, which is the smallest positive value of τ given by Equation (6.15), the conditions for Hopf bifurcation are justified.

Theorem 6.7 Suppose that $E^* = (N_1{}^*(t), N_2{}^*(t), A^*(t))$ exist and the condition in (H_1) are satisfied for the innovation diffusion model (6.1), then the conditions for $E^* = (N_1{}^*(t), N_2{}^*(t), A^*(t))$ to be LAS with τ are

(i) if $\tau \in [0, \tau_0)$, then E^* is LAS;
(ii) if $\tau \geq \tau_0$, the point E^* bifurcates into periodic orbits, i.e., it becomes unstable;
(iii) system (6.1) undergoes Hopf bifurcation at threshold value τ_0 around E^* where

$$\tau_0=\frac{1}{\omega_0}\arctan\left[\frac{\left(P_1\omega_0{}^2-P_3\right)Q_2\omega_0-\left(\omega_0{}^3-P_2\omega_0\right)\left(Q_3-Q_1\omega_0{}^2\right)}{\left(\omega_0{}^3-P_2\omega_0\right)Q_2\omega_0-\left(P_1\omega_0{}^2-P_3\right)\left(Q_3-Q_1\omega_0{}^2\right)}\right].$$

6.7 SENSITIVITY ANALYSIS

Here, the sensitivity analysis at non-negative equilibrium state $E^* = (N_1{}^*(t), N_2{}^*(t), A^*(t))$ of the state variables with respect to model parameters has been found. Sensitivity indices are shown in Table 6.1.

TABLE 6.1

Sensitivity Indices $\Upsilon_{y_j}^{x_i} = \frac{\partial x_i}{\partial y_j} \times \frac{y_j}{x_i}$ of the System (6.1) to the Parameters y_j for the Parameter Values

Parameter (y_j)	Values	Sensitivity of N_1*	Sensitivity of N_2*	Sensitivity of A*
Λ_1	0.25	0.785502	0.0618898	0.210352
Λ_2	0.4	0.0679252	0.297071	1.00969
ξ_1	0.1	−0.237769	0.0534904	0.181804
ξ_2	0.2	0.22026	−0.0366917	−0.124708
α	0.4	−0.146573	−0.641039	0.220041
β	0.3	−0.811102	−0.267232	0.0917292
δ	0.2	0.111869	0.48926	−1.57406
ν	0.01	0.00988932	0.043251	−0.0148462

Definition ([31]) The normalized forward sensitivity index (*S.I.*) of a variable, u, that depends on a parameter p, is defined as: $\Upsilon_p^u = \frac{\partial u}{\partial p} \times \frac{p}{u}$.

The observing points to note that the parameters Λ_1, Λ_2, α, and δ are the most sensitive parameter to A^*.Thus, it can be said that there will be a significant change in the adopter population A^* by small changes in the values of these parameters. However, the recruitment rates Λ_1 and Λ_2 of non-adopter populations N_1 and N_2 has a positive role to play in shaping the dynamics of N_1* and N_2*, and A*.

6.8 NUMERICAL SIMULATIONS

In this section, the analytical results so far obtained will be validated by presenting a numerical example and justify the outcomes of the system (6.1). For this, let us assume a hypothetical set of parametric values, given in the following illustration as

$$\begin{cases} \frac{dN_1}{dt} = 0.25 - 0.1N_1(t) + 0.2N_2(t) - 0.2N_1(t), \\ \frac{dN_2}{dt} = 0.4 + 0.1N_1(t) - 0.2N_2(t) - \big(0.4A(t-\tau) + 0.3\big)N_2(t-\tau) \\ \qquad +0.01A(t) - 0.2N_2(t), \\ \frac{dA}{dt} = \big(0.4A(t-\tau) + 0.3\big)N_2(t-\tau) - (0.2 + 0.01)A(t). \end{cases} \tag{6.20}$$

The detailed explanations of this numerical example can be given by the help of subsequent two cases:

Case I: When $\tau = 0$

The help of MATLAB® software is taken to integrate the model (6.18). By taking distinct sets of different initial values, it has been noticed the system (6.18) is locally asymptotically stable

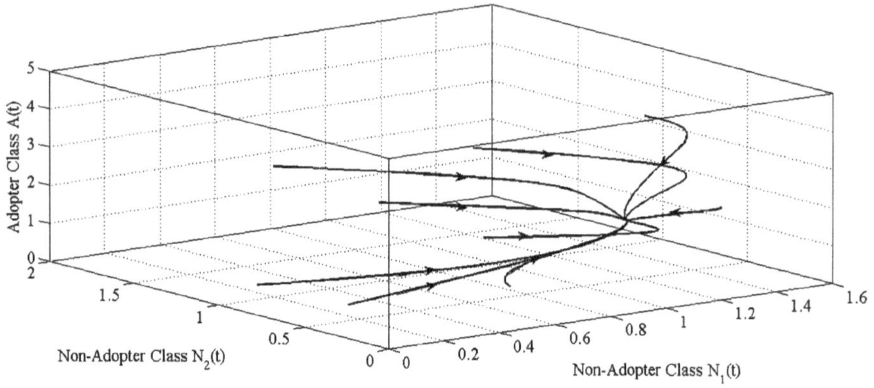

FIGURE 6.1 The solution curves of all the populations are converging to stable equilibrium position $E^0(1.082, 0.3708, 1.799)$.

in the absence of delay parameter τ, and the model (6.18) converges to a stable steady state $E^0(1.082, 0.3708, 1.799)$. It has been depicted by the help of Figure 6.1

Moreover, the Routh–Hurwitz conditions

$$\begin{cases} H_1 : P_1 + Q_1 = 0.241 > 0, P_2 + Q_2 = 1.629 > 0, P_3 + Q_3 = 0.216 > 0, \\ (P_1 + Q_1)(P_2 + Q_2) - (P_3 + Q_3) = 0.1765 > 0. \end{cases}$$

of the (LAS) of the system (6.1) has also been satisfied numerically.

Case II: For $\tau > 0$,

The system (6.18) with value of $\tau > 0$ has again been integrated with initial values (0.3, 0.3, 0.1), and seen that there exists a purely imaginary root $i\omega$, where $\omega = 1.228$ has been calculated and corresponding to this value, the critical value of time delay parameter $\tau_0 = 2.8$ is calculated from Equation (6.14), such that the non-negative steady state E^* remains stable for $0 \le \tau < 2.8$ and is unstable for $\tau \ge 2.8$ (Figure 6.2–6.3). Moreover, the transversality condition

$$\left[\frac{d\mu}{d\tau}\right]_{\tau=\tau_0,\omega=\omega_0} = 1.7728 \neq 0$$

is also verified, which is the prior condition for the existence of Hopf bifurcation. Thus, the system (6.1) producing Hopf-bifurcation via a limit cycle, and has been presented in Figure 6.4 for $\tau = 2.9$. It proves that there exists a threshold limit of delay parameter τ, beyond which the system shows excitability in the form of limit cycle. Also for some more value of parameter, i.e., $\tau = 6.3$, the system (6.1) starts producing irregular oscillations, and then the existence of chaotic attractors around the interior equilibrium E^* for $\tau = 8.4$ has also been shown in the Figures 6.5–6.6. The complexity in the behaviour of model (6.1) with the occurrence of a Hopf bifurcation to irregular oscillations, and then to chaotic situations has been shown by a waveform plot and phase plane diagram in the Figures 6.3–6.6. The analytical and numerical simulations demonstrated that the delay parameter τ in the innovation diffusion process of (6.1) has caused the complexity in the dynamics of the system. The outcomes proved that the

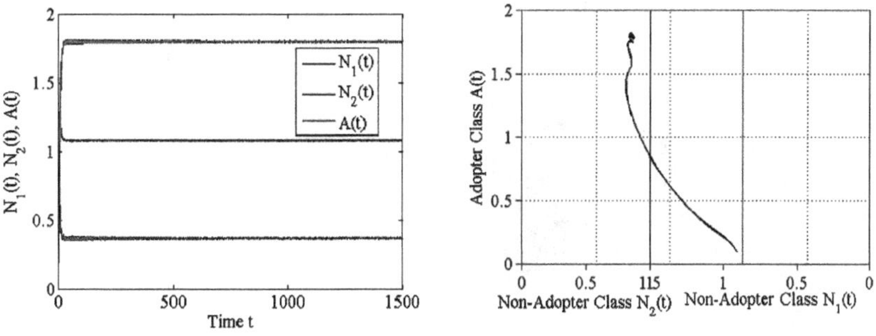

FIGURE 6.2 The time series and phase plots of all Non-adopters and Adopter classes of the system (6.18) are converging to stable interior equilibrium point $E^*(1.077, 0.3742, 1.776)$ at $\tau = 2.77 < 2.8 = \tau_0$)

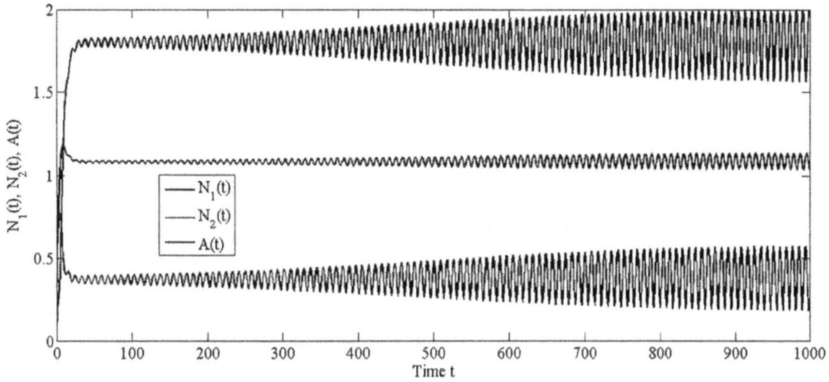

FIGURE 6.3 Time series produces periodic solution trajectories appears around the positive equilibrium $E^*(1.08, 0.3739, 1.782)$ when the value of evaluation period is $\tau = 2.9$.

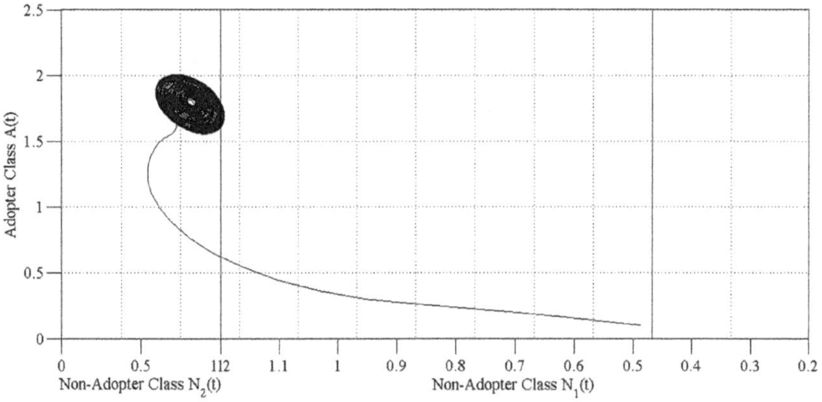

FIGURE 6.4 Phase plane depicts periodic oscillations around the positive equilibrium $E^*(1.08, 0.3739, 1.782)$ when $\tau = 2.9$.

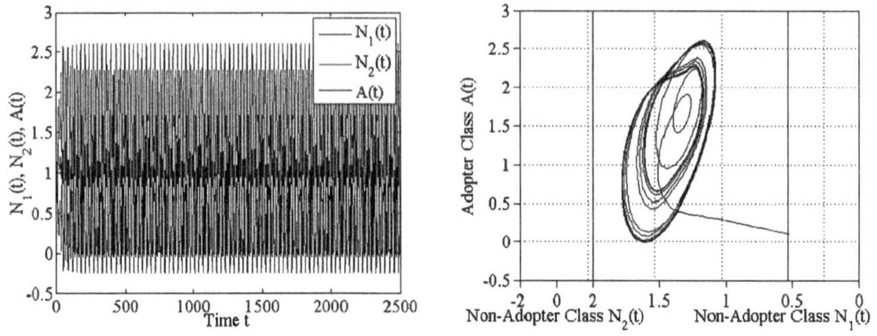

FIGURE 6.5 Irregular oscillations of all the solution curves of populations $N_1(t)$, $N_2(t)$, $A(t)$ are induced for $\tau = 6.3$, which is proving complexities in the innovation diffusion system in (6.18).

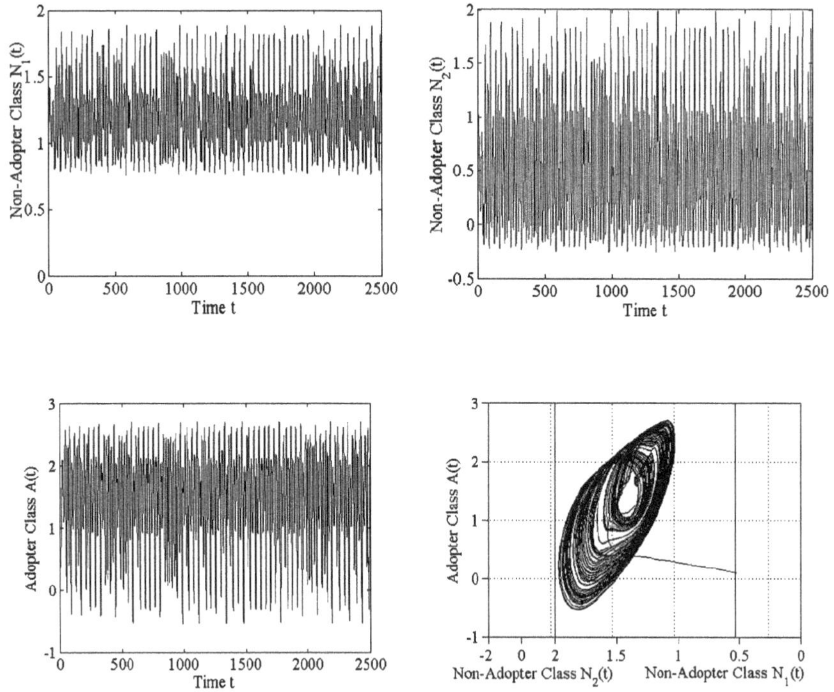

FIGURE 6.6 Chaotic attractors around the interior equilibrium E^* for time delay $\tau = 8.4$ has been observed for all the populations of model system (6.18).

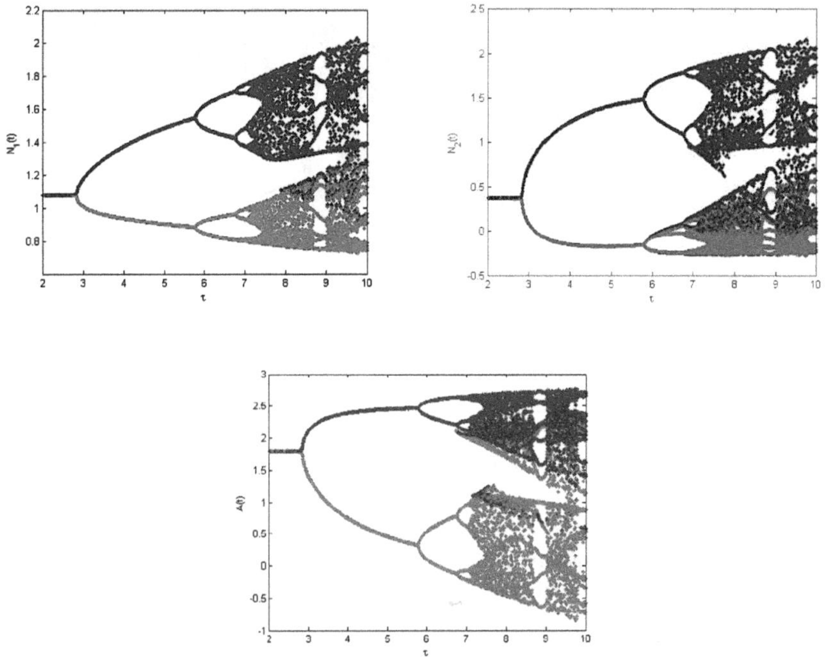

FIGURE 6.7 The bifurcation diagram of the model (6.1) with respect to the bifurcating parameter τ has been shown in this figure, where other parameters are kept same as Figure 6.1. Figure shows that the system becomes chaotic from stable dynamics with the increase in the time delay parameter τ.

time delay parameter is necessary for the innovation to take off in various markets, and it has destabilizing effects on the innovation diffusion model (6.1). It has been transforming the system (6.1) from a locally stable behaviour to a limit cycle via Hopf bifurcation, and then helped it to produce irregular oscillations and finally chaos in the diffusion markets.

6.9 RESULTS AND DISCUSSION

The oscillatory nature in the processes of diffusion of an innovation is a common thing. The model (6.1) has been examined for basic preliminaries, and justified the positivity of the solutions. Also, all the solutions of the model are lying inside the boundary of a defined region. The basic influence number of the system has been analyzed. The locally asymptotic stability behaviour of the model (6.1) is investigated for the adopter-free E^0 and non-negative steady state E^*. The steady-state E^0 is proved as locally stable if the basic influence number is less than one, i.e., $R_A < 1$. It signifies that the diffusion process of the product in the market cannot take off. But if $R_A > 1$, the system (6.1) becomes unstable.

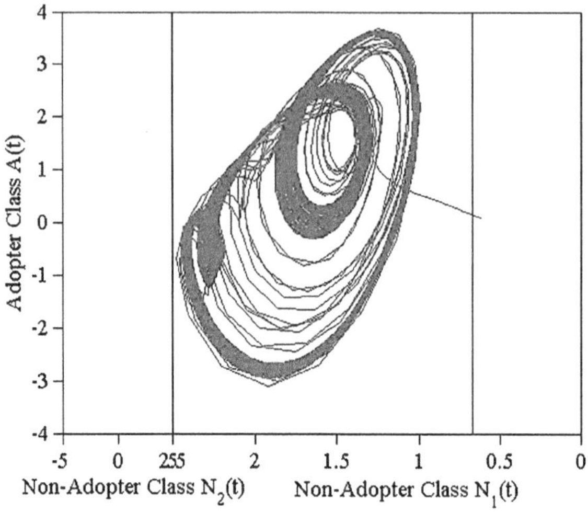

FIGURE 6.8 The effects of increase in advertisements on model (6.1) have been shown in this figure, where other parameters are kept same as Figure 6.1. Figure showed that the system becomes complex for $\beta = 0.8$ with time delay parameter $\tau = 2.9$.

The system (6.1) established locally stable behaviour without any evaluation period. But the delay parameter τ in the system (6.1) pushes the system into Hopf bifurcation, i.e., small amplitude periodic oscillations have been anticipated around the nonnegative steady state E^*. This means that as and when the value of τ crossed over the threshold value $\tau^*= 2.8$, the system is proven to be admitting stability switch, and showed that there exists a limit cycle. More stable periodic oscillations of the model (6.1) around E^* for non-adopter and adopter populations at $\tau = 2.9$ are shown in Figure 6.4. The model system (6.1) showed irregular oscillations of solution curves are induced at $\tau = 6.3$ in Figure 6.5. The chaotic attractors around E^* for time delay $\tau = 8.4$ have also been shown in Figure 6.6 for all population classes. With the help of the bifurcation diagram of the system (6.1), it has been exhibited that the system becomes more complex (chaotic) from stable dynamics with a little increase in the value of parameter τ. In Figure 6.8, it has been observed that a small increase in the advertisements give rise to complexities in the system (6.1). The sensitivity analysis about the most sensitive parameters have also been executed, and presented in Table 6.1 by computing sensitivity indices of adopter population A^*.

6.10 CONCLUSION

Here, in this chapter, an economic group model with the delay in adoption of an innovation has been given, and the dynamical behaviour of the continuous diffusion model with advertisements as well as interpersonal communications has been examined. The analytical and numerical results are well in agreement of the fact that the delay in the word of mouths effects between non-adopters population $N_2^*(t)$ and the adopter's population $A^*(t)$ has produced small periodic oscillations in the system, and then complex situations in various markets. Similarly, a small increase in the external advertisements can also create irregularities in the innovation diffusion markets. So, the important aspect of this work is to apply the optimized extents of advertisements as well as interpersonal communications which are helpful in the diffusion processes to take off in the various markets.

REFERENCES

[1] V. Mahajan, E. Muller, and F. Bass, New product diffusion models in marketing: A review and directions for research. *Journal of Marketing*, 54, 1–26, 1990.

[2] E. M. Rogers. *Diffusion of innovations*, The Free Press, New York, 1995.

[3] F. M. Bass, A new product growth model for consumer durables. *Management Science*, 15, 215–227, 1969.

[4] M. N. Sharif, and C. Kabir, A generalized model for forecasting technology substitution. *Technological Forecasting and Social Change*, 8, 353–361, 1976.

[5] W. Wang, P. Fergola, and C. Tenneriello, Innovation diffusion model in patch environment. *Applied Mathematics and Computation*, 134(1), 51–67, 2003.

[6] W. Wang, P. Fergola, S. Lombardo, and G. Mulone, Mathematical models of innovation diffusion with stage structure. *Applied Mathematical Modelling*, 30(1), 129–146, 2006.

[7] Y. Yu, and W. Wang, Stability of innovation diffusion model with nonlinear acceptance. *Acta Mathematica Scientia*, 27(3), 645–655, 2007.

[8] J. B. Shukla, H. Kushwah, K. Agrawal, and A. Shukla, Modeling the effects of variable external influences and demographic processes on innovation diffusion. *Nonlinear Analysis: Real World Applications*, 13(1), 186–196, 2012.

[9] C. Tenneriello, P. Fergola, M. Zhien, and W. Wang, Stability of competitive innovation diffusion model. *Ricerche di Matematica*, 51(2), 185–200, 2002.

[10] F. Centrone, A. Goia, and E. Salinelli, Demographic processes in a model of innovation diffusion with dynamic market. *Technological Forecasting and Social Change*, 74(3), 247–266, 2007.

[11] W. M. Cohen, and D. A. Levinthal, Absorptive capacity: A new perspective on learning and innovation. *Administrative Science Quarterly*, 35, 128–152, 1990.

[12] N. Boccara, H. Fuks, and S. Geurten, A new class of automata networks. *Physica D*, 103, 145–154, 1997.

[13] V. Mahajan, E. Muller, and R. A. Kerin, Introduction strategy for new products with positive and negative word-of-mouth. *Management Science*, 30, 1389–1404, 1984.

[14] R. Guseo, and M. Guidolin, Cellular automata and riccati equation models for diffusion of innovations. *Statistical Methods and Applications*, 17(3), 291–308, 2008.

[15] V. Mahajan, and S. Sharma, A simple algebraic estimation procedure for innovation diffusion models of new product acceptance. *Technological Forecasting and Social Change*, 30, 331–345, 1986.

[16] Xie J., X. M. Song, M. Sirbu, and Q. Wang, Kalman filter estimation of new product diffusion models. *Journal of Marketing Research*, XXXIV, 378–393, 1997.

[17] V. Srinivasan, and C. H. Mason, Nonlinear least squares estimation of new product diffusion models. *Marketing Science*, 5(2), 169–178, 1986.

[18] D. C. Schmittlein, and V. Mahajan, Maximum likelihood estimation for an innovation diffusion model of new product acceptance. *Marketing Science*, 1(1), 57–78, 1982.

[19] C. Kennedy, *Guide to the Management Gurus: Shortcuts to the Ideas of Leading Management Thinkers*, Business Books, London, 1991.

[20] H. Singh, J. Dhar, H. S. Bhatti, and S. Chandok, An epedemic model of childhood disease dynamics with maturation delay and latent period of infection. *Modeling Earth Systems and Environment (MESE)*, 2(2), 1–8, 2016.

[21] R. Kumar, A. K. Sharma, and K. Agnihotri, Stability and bifurcation analysis of a delayed innovation diffusion model. *Acta Mathematica Scientia*, 38(2), 709–732, 2018.
[22] J. C. Cortes, I. C. Lombana, and R. J. Villanueva, Age-Structured mathematical modeling approach to short-term diffusion of electronic commerce in Spain. *Mathematical and Computer Modelling*, 52, 1045–1051, 2010.
[23] R. Nie, K. M. Qian, and D. H. Pan, The innovative technology diffusion models and their stability analysis base on logistic equation. *Journal of Industrial Engineering and Management*, 20(1), 41–45, 2006.
[24] B. Zhang, L. Fang, and R. B. Zhang, A dynamic diffusion model of competitive multi-innovations and its application. *Technology Economics*, 27(9), 5–9, 2008.
[25] J. Dhar, M. Tyagi, and P. Sinha, Three simultaneous innovations interrelationships: An adopter dynamics model. *International Journal of Modeling, Simulation, and Scientific Computing*, 6, 3, 2015.
[26] J. K. Hale, *Functional Differential Equations*, Springer, New York, 1971.
[27] W. G. Aiello, and H. Freedman, A time-delay model of single-species growth with stage structure. *Mathematical Biosciences*, 101(2), 139–153, 1990.
[28] G. Birkhoff, and G. Rota, *Ordinary Differential Equations*, Ginn, Boston, 1989.
[29] P. V. Driwssche, and J. Watmough, Reproduction numbers and sub-threshold endemic equilibria for compartmental models of disease transmission. *Mathematical Biosciences*, 180, 29–48, 2002.
[30] S. Boonrangsiman, K. Bunwong, and E. J. Moore, A bifurcation path to chaos in a time-delay fisheries predator–prey model with prey consumption by immature and mature predators. *Mathematics and Computers in Simulation*, 124, 16–29, 2016.
[31] G. P. Sahu, and J. Dhar, Dynamics of an seqihrs epidemic model with media coverage, quarantine and isolation in a community with pre-existing immunity. *Journal of Mathematical Analysis and Applications*, 421(2), 1651–1672, 2015.

7 Stochastic Analysis of a Priority-Based Warm Standby System Working under k-out-of-n *G Policy Using Multi-Dimensional Repair*

Praveen Kumar Poonia and Anu Sirohi

CONTENTS

DOI: 10.1201/9781003089636-7

7.1 INTRODUCTION

We have a broad range of complex systems in which the failure of any component will reduce the performance of the overall system or the total failure of the system and thereby reduce the reliability of the system. Experience with low reliability is still a major concern in the scientific world and depends on more systematic approaches of reliability theory. Numerous methods are available to improve reliability, from which standby redundancy (sometimes referred to as backup redundancy) is widely used and is able to improve the system availability and reliability. Furthermore, a number of applications of standby systems are available in different dynamic engineering systems. There are three forms of standby available: (i) Cold standby, where the standby unit is used if the main or operating unit fails and has a zero failure rate when in standby; (ii) Hot standby, where the standby unit has the same failure rate as the operating unit; and (iii) Warm standby, where the standby unit runs in the surroundings of the operating unit. This is an intermediate condition and the failure of unit is likely in this situation at a reduced failure rate as compared to the operating unit. In the real world, we come across a number of dynamical engineering systems with standby units, such as the power generation and transmission system, truck/bus wheel assembly, housing inverters, etc. In these systems, while the operating unit fails, the load is moved spontaneously to the standby unit(s) in the case of a warm standby. The most common form of redundancy is k-out-of-n, which is well known to all and widely acknowledged by many industries and establishments. Several articles have been published over the past two decades with reference to reliability and availability of k-out-of-n systems. In specific, authors consider warm standby systems [1], comparison of combined and consecutive units [2], multistate systems under various conditions [3], an approximation algorithm to calculate time to failure using shared standby elements [4], a reliability formula for consecutive repairable systems [5], reliability of k-out-of-r from n with imperfect fault coverage [6], generalized block replacement policy using the restriction on number of failed components and hazard costs [7], consecutive system with non-identical lifetimes for components [8], standby redundancy techniques with statistically independent units [9], non-repairable dynamic systems with sequence-dependent failures [10], steady-state availability of cold standby repairable system [11], and many more.

Most of the equipment and machines are made up of a variety of components. The reliability of each component and the configuration of the system consisting of these components assess the system reliability. Any of them could fail, and this increases the chances that the whole system will fail. Product design, manufacture, and maintenance all have an effect on reliability. Improving reliability is largely the domain of design. Adding redundancy increases the cost and complexity of the design of the system. Although there are different approaches, strategies, and terminologies for the implementation of redundancy, it is used in the form of identical components connected in such a manner that, when one component fails, the others will keep the system functioning. One common practice for increasing reliability is that components are mounted in parallel. As long as one component operates, the system operates. The system fails when all components in parallel fail. Several authors, such as Kumar [12], Ram [13], Singh [14], Tiwari [15] and Yusuf [16], have analyzed two unit warm standby parallel systems under various assumptions and calculated all major reliability indices, including reliability, availability, mean time to failure (MTTF) and expected cost benefit, assuming that the working time and repair time of each component is compatible with exponential distribution. In addition, every unit/component works as a new unit after repair. Nevertheless, the investigation of three or more standby unit systems attracted very little, though very considerable, possibly some unidentified technological bumps while investigating them. Some researchers have studied these systems in which they assume the failure rate and repair rate to be distributed exponentially; if both distributions were taken as general ones, however, the problem would be transformed into an out-of-control mode. Srinivasan et al. [17] have tested a three-unit warm standby-redundant system and showed that it has the potential for a comprehensive analysis. During the study, they found various renewal points that made the review feasible. These imbedded renewal points allowed authors to find the reliability characteristics. Munjal and Singh [18] studied 2-out-of-3: G in parallel and five another type of components in series configuration and evaluated reliability characteristics using supplementary variable technique. Goyal et al. [19] tested a three-unit redundant series system using k-out-of-n: G policy and carried out a sensitivity analysis for the system. In this, the authors consider three subsystems, A, B, and C such that the first subsystem, A, has n units arranged in parallel, the second subsystem, B, has two subsystems, X and Y, in parallel, where X functions under 1-out-of-n: F. The system is deeply studied and assessed for availability and reliability. Gupta et al. [20] examined three non-identical units' parallel system, where the third unit was positioned in cold standby mode under 1-out-of-3: G configuration. In this chapter, for a partial failure mode, the authors employed general repair, while for complete failure, the authors demonstrated an immediate need for a certain special facility capable of enhancing system reliability and availability, for which they include Gumbel-Hougaard copula repair [21]. With reference to this strategy authors, including Singh et al. [22], have premeditated cost analysis of a computer network that includes three computer labs in parallel configuration connected in series through a server under the concept of k-out-of-n: G policy. The authors used copula repair and show beyond doubt that copula repair is far better as compared to general repair in a complete failed system. Gehlot et al. [23] have carried out performance analysis of a complex system divided into two subsystems in series configuration. The first subsystem has three units in operation using policy 2-out-of-3: F, and the second subsystem has two units operational using policy 1-out-of-2: G. The failure rates of units are constant and expected to go along with exponential distribution and their repair backing two types of repairs, namely general repair and copula repair. Lado et al. [24] analyzed two subsystems connected in series configuration and the whole system is functioned by a human operator and prove that copula repair is more worthy on general repair. Recently, Raghav et al. [25] and Singh et al. [26] have examined the cost analysis of a complex system comprising two subsystems with five identical units and two identical /non-identical units, respectively, in a series configuration with an imperfect switch. A sensitive analysis was carried out by the authors and the MTTF was calculated, using copula repair for complete failed units. Poonia and Sirohi [27] considered a series configuration with three subsystems having three, two and one units respectively, working under the k-out-of-n: G policy. The authors have considered all failure rates constant and a two repair policy, viz. general repair and copula repair, with the deliberate failure of the system. A repairable warm standby k-out-of-n: G and 2-out-of-4: G systems in series under catastrophic failure and a switching device was recently studied by Poonia et al. [28] using copula repair. This model was developed by considering n-k+1 states in subsystem-1 in such a way that it formed a finite series during solution, unlike in the past. In addition, availability and profit are higher when repair supports copula family distribution. Accordingly, in this chapter the authors encouraged the scientific community to implement multidimensional copula repairs.

7.2 MODEL DESCRIPTION AND NOTATIONS

7.2.1 System Description

In the past, researchers have been very interested in the analysis of repairable complex systems by computing the performance under different failure and repair policies, but very few have considered priorities in available units. Nevertheless, some study is needed in order to increase supporting units and considering catastrophic failure. Keeping this need in mind, in this chapter we considered a 3-out-of-5: G warm standby system consisting of five units in parallel configuration under catastrophic failure. All the five units are divided into two categories: 3 units as type-A and 2 units as type-B units. Type-A units have preference in operation and repair over type-B units, which means type-A components are preferred to perform where available. One repairperson is available full time to restore the system. We have considered a catastrophic failure resulting from abrupt changes in weather conditions or man-made disruptions. The failure rates of units are constant and are subject to exponential distribution, but their repair supports general repair and copula repair. The system we considered would fail only if any three units failed with preference to type-A units. Over the process of the operation, the system would be in any of the three states: Good operation: all units are in functioning mode, partial failure: running under its design conditions and complete failure: the system ceases to operate. Using a supplementary variable technique, Laplace transforms and copula, we evaluate various characteristics such as transition state probabilities, availability, reliability, MTTF and profit analysis. The divisions of this chapter are planned as follows. Section 7.2 of the manuscript includes system description, assumptions, and state description. Section 7.3 consists of system configuration and transition diagram. The state explanation is provided in Section 7.4. Section 7.5 introduces the mathematical modeling using differential equations and Laplace transforms. The findings for the different output measures of the system are simulated by considering a variety of special cases mentioned in Section 7.6. Using the graphs, the final remarks on our findings with interpretations are given in Section 7.7. MAPLE computer software is used for numerical analysis of reliability, availability, MTTF and cost analysis.

7.2.2 Expectations Aimed for the Model

The following assumptions for the priority-based model under study are proposed:

1. This is a priority-based, warm standby system operating under a 3-out-of-5: G policy successfully if three or more units are in good working condition and ready for use.
2. Initially, all type-A units are functioning while type-B units are in warm standby mode. Type-A units are much more preferred for operation and repair than type-B units.
3. Type-B units will come from warm standby mode when any type-A unit fails and cannot function normally. After repair, it becomes operational again and replaces one of type-B. The replaced unit would be in standby again until the system is able to work normally.
4. The failure rate of each type-A unit is λ_1 and each type-B unit is λ_2, while it is lower for type-A units ($\lambda_1 < \lambda_2$) as compared to type-B units.
5. One repairperson is always available and may be called as the system progresses to partially or completely failed state.
6. All the units of type-A and type-B are in parallel mode. If a type-A unit fails, the type-B unit will operate automatically.
7. All failure rates are taken as constant and are to be accompanied by an exponential distribution.
8. Copula repair is used while system is entering a complete failure mode to restore it.

7.2.3 Notations

s, t	Laplace transform and time scale variable.
λ_1/λ_2	Failure rate of type-A/type-B units.
λ_C	Failure rate due to catastrophic failure mode.
$\phi_1(x)/\phi_2(x)$	Repair rate of type-A/type-B units.
$\mu_0(x)$	Repair rate due to total failure.
$P_0(t)$	The state transition probability that the system is in state S_0.
$P_i(x, t)$	The probability that the system is in state S_i and under repair where repair variable is x and time variable is t.
$E_p(t)$	Expected profit incurred during time t.
K_1, K_2	Revenue generated and service cost per unit time respectively.
$\mu_0(x)$	Joint probability function from complete failed state S_i to S_0.

7.3 SYSTEM CONFIGURATION AND STATE TRANSITION DIAGRAM

System configuration and state transition diagrams are presented in Figure 7.1 and 7.2 respectively. In the transition diagram, S_0 is the flawless state, S_1, S_2, S_3, S_4 and S_5 partially failed states and S_6, S_7, S_8, S_9 and S_{10} are completely failed states. After the failure of maximum two type-A and type-B units, the transitions progressed to

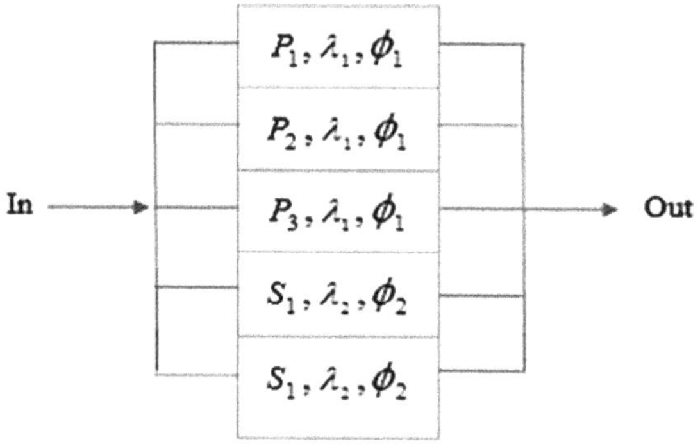

FIGURE 7.1 System configuration.

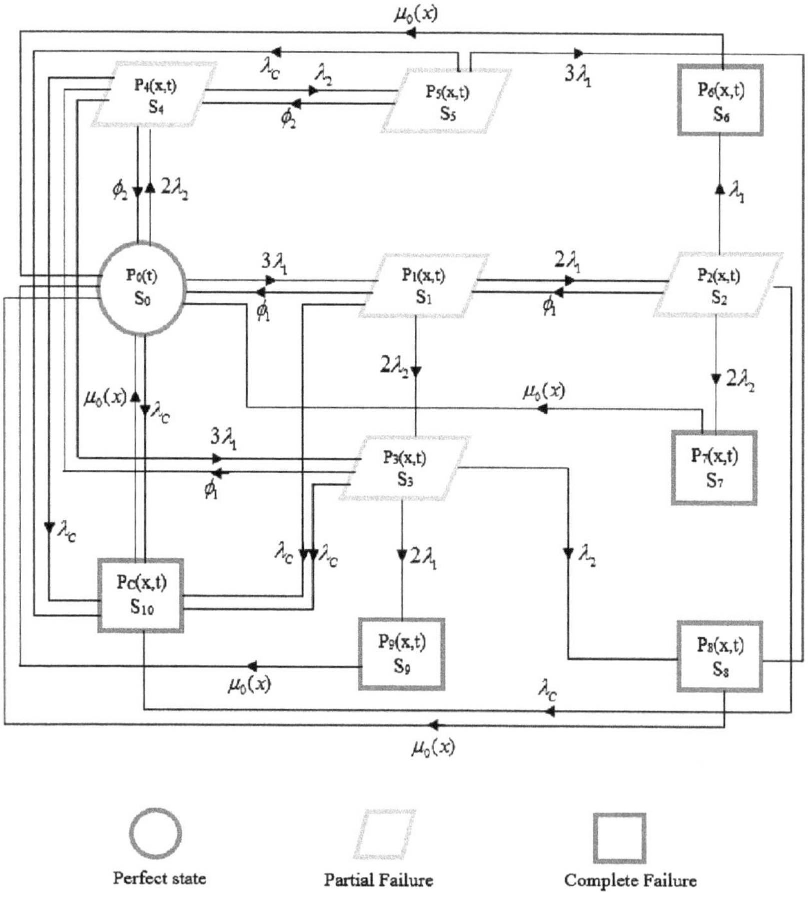

FIGURE 7.2 State transition diagram of the model.

partially failed states S_1, S_2, S_3, S_4 and S_5. The state S_6, S_7, S_8 and S_9 are complete failed states due to failure of any three units. S_{10} is completely failed state owing to catastrophic failure.

7.4 STATE EXPLANATION

The explanation of several feasible states of the model after failing the units of type-A and type-B, including catastrophic failure, is shown in Table 7.1. The states $\{S_0, S_1, S_2, S_3, S_4$ and $S_5\}$ are functioning states and $\{S_6, S_7, S_8, S_9$ and $S_{10}\}$ are non-functioning states of the system.

All the degraded/partially failed states are treated via general repair, while complete failed states are treated via Gumbel–Hougaard family copula repair.

7.5 MATHEMATICAL FORMULARIZATIONS FOR THE MODEL

With the help of probability of considerations and continuity arguments, we try to get difference-differential equations associated with the current mathematical model.

$$\left[\frac{\partial}{\partial t}+3\lambda_1+2\lambda_2+\lambda_C\right]P_0(t)=\int_0^\infty \phi_1(x)P_1(x,t)dx+\int_0^\infty \phi_2(x)P_4(x,t)dx+$$
$$\sum_k\int_0^\infty \exp\left[x^\theta+\{\log\phi(x)\}^\theta\right]^{1/\theta}P_k(x,t)dx\{k=6,7,8,9,10\} \tag{7.1}$$

$$\left[\frac{\partial}{\partial t}+\frac{\partial}{\partial x}+2\lambda_1+2\lambda_2+\lambda_C+\phi_1(x)\right]P_1(x,t)=0 \tag{7.2}$$

$$\left[\frac{\partial}{\partial t}+\frac{\partial}{\partial x}+\lambda_1+2\lambda_2+\lambda_C+\phi_1(x)\right]P_2(x,t)=0 \tag{7.3}$$

TABLE 7.1
State Representation of the Model

State	State Description	No. of good units	State	State Description	No. of good units
S_0 (H)	$P_{1o}, P_{2o}, P_{3o}, S_{1s}, S_{2s}$	5	S_6 (F)	$P_{1r}, P_{2wr}, P_{3wr}, S_{1o}, S_{2o}$	2
S_1 (D)	$P_{1r}, P_{2o}, P_{3o}, S_{1o}, S_{2s}$	4	S_7 (F)	$P_{1r}, P_{2wr}, P_{3o}, S_{1wr}, S_{2o}$	2
S_2 (D)	$P_{1r}, P_{2wr}, P_{3o}, S_{1o}, S_{2o}$	3	S_8 (F)	$P_{1r}, P_{2o}, P_{3o}, S_{1wr}, S_{2wr}$	2
S_3 (D)	$P_{1r}, P_{2o}, P_{3o}, S_{1wr}, S_{2o}$	3	S_9 (F)	$P_{1r}, P_{2wr}, P_{3o}, S_{1wr}, S_{2o}$	2
S_4 (D)	$P_{1o}, P_{2o}, P_{3o}, S_{1r}, S_{2s}$	4	S_{10} (F)	Catastrophic Failure	-
S_5 (D)	$P_{1o}, P_{2o}, P_{3o}, S_{1r}, S_{2wr}$	3			

H-Healthy State, D-Degraded/Partially Failed State, F-Complete Failed State.
P-priority unit, S-supporting unit, o-in operation, s-standby, r-under repair, and wr-waiting for repair.

$$\left[\frac{\partial}{\partial t}+\frac{\partial}{\partial x}+2\lambda_1+\lambda_2+\lambda_C+\phi_1(x)\right]P_3(x,t)=0 \tag{7.4}$$

$$\left[\frac{\partial}{\partial t}+\frac{\partial}{\partial x}+3\lambda_1+\lambda_2+\lambda_C+\phi_2(x)\right]P_4(x,t)=0 \tag{7.5}$$

$$\left[\frac{\partial}{\partial t}+\frac{\partial}{\partial x}+3\lambda_1+\lambda_C+\phi_2(x)\right]P_5(x,t)=0 \tag{7.6}$$

$$\left[\frac{\partial}{\partial t}+\frac{\partial}{\partial x}+\exp\left[x^\theta+\{\log\phi(x)\}^\theta\right]^{1/\theta}\right]P_k(x,t)=0\{k=6,7,8,9,10\} \tag{7.7}$$

Boundary conditions

$$P_1(0,t)=3\lambda_1P_0(t) \tag{7.8}$$

$$P_2(0,t)=2\lambda_1P_1(0,t)=6\lambda_1^2P_0(t) \tag{7.9}$$

$$P_4(0,t)=2\lambda_2P_0(t) \tag{7.10}$$

$$P_3(0,t)=3\lambda_1P_4(0,t)+2\lambda_2P_1(0,t)=12\lambda_1\lambda_2P_0(t) \tag{7.11}$$

$$P_5(0,t)=\lambda_2P_4(0,t)=2\lambda_2^2P_0(t) \tag{7.12}$$

$$P_6(0,t)=\lambda_1P_2(0,t)=6\lambda_1^3P_0(t) \tag{7.13}$$

$$P_7(0,t)=2\lambda_2P_2(0,t)=12\lambda_1^2\lambda_2P_0(t) \tag{7.14}$$

$$P_8(0,t)=\lambda_2P_3(0,t)+3\lambda_1P_5(0,t)=18\lambda_1\lambda_2^2P_0(t) \tag{7.15}$$

$$P_9(0,t)=2\lambda_1P_3(0,t)=24\lambda_1^2\lambda_2P_0(t) \tag{7.16}$$

$$P_C(0,t)=\lambda_C\left[P_0(t)+P_1(0,t)+P_2(0,t)+P_3(0,t)+P_4(0,t)+P_5(0,t)\right] \tag{7.17}$$

Initial conditions

$$P_0(0)=1,\ \text{and rest probabilities are zero at}\quad t=0 \tag{7.18}$$

For the solution, let us take Laplace transformation of Equations (7.1) to (7.17) with the help of initial condition (7.18), we get

$$\begin{aligned}\left[s+3\lambda_1+2\lambda_2+\lambda_C\right]\bar{P}_0(s)&=1+\int_0^\infty\phi_1(x)\bar{P}_1(x,s)dx+\int_0^\infty\phi_2(x)\bar{P}_4(x,s)dx\\&+\sum_k\int_0^\infty\exp\left[x^\theta+\{\log\phi(x)\}^\theta\right]^{1/\theta}\bar{P}_k(x,s)dx\{k=6,7,8,9,10\}\end{aligned} \tag{7.19}$$

$$\left[s+\frac{\partial}{\partial x}+2\lambda_1+2\lambda_2+\lambda_C+\phi_1(x)\right]\bar{P}_1(x,s)=0 \tag{7.20}$$

$$\left[s+\frac{\partial}{\partial x}+\lambda_1+2\lambda_2+\lambda_C+\phi_1(x)\right]\bar{P}_2(x,s)=0 \tag{7.21}$$

$$\left[s+\frac{\partial}{\partial x}+2\lambda_1+\lambda_2+\lambda_C+\phi_1(x)\right]\bar{P}_3(x,s)=0 \tag{7.22}$$

$$\left[s+\frac{\partial}{\partial x}+3\lambda_1+\lambda_2+\lambda_C+\phi_2(x)\right]\bar{P}_4(x,s)=0 \tag{7.23}$$

$$\left[s+\frac{\partial}{\partial x}+3\lambda_1+\lambda_C+\phi_2(x)\right]\bar{P}_5(x,s)=0 \tag{7.24}$$

$$\left[s+\frac{\partial}{\partial x}+\exp\left[x^{\theta}+\{\log\phi(x)\}^{\theta}\right]^{1/\theta}\right]\bar{P}_k(x,s)=0\{k=6,7,8,9,10\} \tag{7.25}$$

Boundary conditions

$$\bar{P}_1(0,s)=3\lambda_1\bar{P}_0(s) \tag{7.26}$$

$$\bar{P}_2(0,s)=2\lambda_1\bar{P}_1(0,s)=6\lambda_1^2\bar{P}_0(s) \tag{7.27}$$

$$\bar{P}_4(0,s)=2\lambda_2\bar{P}_0(s) \tag{7.28}$$

$$\bar{P}_3(0,s)=3\lambda_1\bar{P}_4(0,s)+2\lambda_2\bar{P}_1(0,s)=12\lambda_1\lambda_2\bar{P}_0(s) \tag{7.29}$$

$$\bar{P}_5(0,s)=\lambda_2\bar{P}_4(0,s)=2\lambda_2^2\bar{P}_0(s) \tag{7.30}$$

$$\bar{P}_6(0,s)=\lambda_1\bar{P}_2(0,s)=6\lambda_1^3\bar{P}_0(s) \tag{7.31}$$

$$\bar{P}_7(0,s)=2\lambda_2\bar{P}_2(0,s)=12\lambda_1^2\lambda_2\bar{P}_0(s) \tag{7.32}$$

$$\bar{P}_8(0,s)=\lambda_2\bar{P}_3(0,s)+3\lambda_1\bar{P}_5(0,s)=18\lambda_1\lambda_2^2\bar{P}_0(s) \tag{7.33}$$

$$\bar{P}_9(0,s)=2\lambda_1\bar{P}_3(0,s)=24\lambda_1^2\lambda_2\bar{P}_0(s) \tag{7.34}$$

$$\bar{P}_C(0,s)=\lambda_C\left[1+3\lambda_1+2\lambda_2+6\lambda_1^2+2\lambda_2^2+12\lambda_1\lambda_2\right]\bar{P}_0(s) \tag{7.35}$$

Laplace transformation of boundary conditions after repair.

$$\bar{P}_1(0,s) = 3\lambda_1\bar{P}_0(s) + \int_0^\infty \phi_1(x)\bar{P}_2(x,s)\,dx$$

$$= 3\lambda_1\bar{P}_0(s) + \int_0^\infty \phi_1(x)\left(e^{-(s+\lambda_1+2\lambda_2+\lambda_C)x-\int_0^x \phi_1(x)dx}\right)\bar{P}_2(0,s)\,dx$$

$$= 3\lambda_1\bar{P}_0(s) + \bar{S}_{\phi_1}(s+\lambda_1+2\lambda_2+\lambda_C)\bar{P}_2(0,s)$$

$$= 3\lambda_1\bar{P}_0(s) + \bar{S}_{\phi_1}(s+\lambda_1+2\lambda_2+\lambda_C)2\lambda_1\bar{P}_1(0,s)$$

$$\bar{P}_1(0,s) = \frac{3\lambda_1}{1-2\lambda_1\bar{S}_{\phi_1}(s+\lambda_1+2\lambda_2+\lambda_C)}\bar{P}_0(s) = \frac{3\lambda_1}{1-2\lambda_1 P}\bar{P}_0(s) \tag{7.36}$$

and

$$\bar{P}_4(0,s) = 2\lambda_2\bar{P}_0(s) + \int_0^\infty \phi_1(x)\bar{P}_3(x,s)\,dx + \int_0^\infty \phi_2(x)\bar{P}_5(x,s)\,dx$$
$$= \frac{2\lambda_2 - 2\lambda_1\lambda_2(2P-3Q)}{(1-2\lambda_1 P)(1-3\lambda_1 P-\lambda_2 R)}\bar{P}_0(s) \tag{7.37}$$

No change in other boundary conditions. Solving the above equations using all the generated boundary conditions with and without repair, we receive

$$\bar{P}_0(s) = \frac{1}{D(s)} \tag{7.38}$$

$$\bar{P}_1(s) = \frac{3\lambda_1}{D(s)}\frac{(1-S)}{(s+2\lambda_1+2\lambda_2+\lambda_C)(1-2\lambda_1 P)} \tag{7.39}$$

$$\bar{P}_2(s) = \frac{6\lambda_1^2}{D(s)}\frac{(1-P)}{(s+\lambda_1+2\lambda_2+\lambda_C)} \tag{7.40}$$

$$\bar{P}_3(s) = \frac{12\lambda_1\lambda_2}{D(s)}\frac{(1-Q)}{(s+2\lambda_1+\lambda_2+\lambda_C)} \tag{7.41}$$

$$\bar{P}_4(s) = \frac{2\lambda_2 - 2\lambda_1\lambda_2(2P-3Q)}{D(s)}\frac{(1-T)}{(s+3\lambda_1+\lambda_2+\lambda_C)(1-2\lambda_1 P)(1-3\lambda_1 P-\lambda_2 R)} \tag{7.42}$$

$$\bar{P}_5(s) = \frac{2\lambda_2^2}{D(s)}\frac{(1-R)}{(s+3\lambda_1+\lambda_C)} \tag{7.43}$$

$$\bar{P}_6(s) = \frac{6\lambda_1^3}{D(s)}\frac{(1-U)}{s} \tag{7.44}$$

$$\bar{P}_7(s) = \frac{12\lambda_1^2\lambda_2}{D(s)}\frac{(1-U)}{s} \tag{7.45}$$

$$\bar{P}_8(s) = \frac{18\lambda_1\lambda_2^2}{D(s)}\frac{(1-U)}{s} \tag{7.46}$$

$$\bar{P}_9(s) = \frac{24\lambda_1^2\lambda_2}{D(s)}\frac{(1-U)}{s} \tag{7.47}$$

$$\bar{P}_C(s) = \frac{\lambda_C\left(1+3\lambda_1+2\lambda_2+6\lambda_1^2+2\lambda_2^2+12\lambda_1\lambda_2\right)}{D(s)}\frac{(1-U)}{s} \tag{7.48}$$

where $D(s) = s+3\lambda_1+2\lambda_2+\lambda_C - \frac{3\lambda_1 S}{1-2\lambda_1 P} - \frac{T}{1-2\lambda_1 P}\frac{2\lambda_2 - 2\lambda_1\lambda_2(2P-3Q)}{1-3\lambda_1 P - \lambda_2 R} - U\left\{6\lambda_1^3+36\lambda_1^2\lambda_2+18\lambda_1\lambda_2^2+\lambda_C\left(1+3\lambda_1+2\lambda_2+6\lambda_1^2+2\lambda_2^2+12\lambda_1\lambda_2\right)\right\}$

and $P = \bar{S}_{\phi_1}(s+\lambda_1+2\lambda_2+\lambda_C) = \frac{\phi_1}{s+\lambda_1+2\lambda_2+\lambda_C+\phi_1}$

$$Q = \bar{S}_{\phi_1}(s+2\lambda_1+\lambda_2+\lambda_C) = \frac{\phi_1}{s+2\lambda_1+\lambda_2+\lambda_C+\phi_1}$$

$$R = \bar{S}_{\phi_2}(s+3\lambda_1+\lambda_C) = \frac{\phi_2}{s+3\lambda_1+\lambda_C+\phi_2}$$

$$S = \bar{S}_{\phi_1}(s+2\lambda_1+2\lambda_2+\lambda_C) = \frac{\phi_1}{s+2\lambda_1+2\lambda_2+\lambda_C+\phi_1}$$

$$T = \bar{S}_{\phi_2}(s+3\lambda_1+\lambda_2+\lambda_C) = \frac{\phi_2}{s+3\lambda_1+\lambda_2+\lambda_C+\phi_2}$$

$$U = \bar{S}_{\mu_0}(s) = \frac{\mu_0}{s+\mu_0}$$

Let $\bar{P}_{up}(s)$ be the sum of Laplace transformations in upstate and $\bar{P}_{down}(s)$ be the sum of Laplace transformations in downstate, then we have

$$\bar{P}_{up}(s) = \bar{P}_0(s)+\bar{P}_1(s)+\bar{P}_2(s)+\bar{P}_3(s)+\bar{P}_4(s)+\bar{P}_5(s)$$

$$= \frac{1}{D(s)}\left[\begin{array}{l} 1+\frac{3\lambda_1(1-S)}{(s+2\lambda_1+2\lambda_2+\lambda_C)(1-2\lambda_1 P)} \\ +\frac{6\lambda_1^2(1-P)}{(s+\lambda_1+2\lambda_2+\lambda_C)}+\frac{12\lambda_1\lambda_2(1-Q)}{(s+2\lambda_1+\lambda_2+\lambda_C)} \\ +\frac{(1-T)(2\lambda_2-2\lambda_1\lambda_2(2P-3Q))}{(s+3\lambda_1+\lambda_2+\lambda_C)(1-2\lambda_1 P)(1-3\lambda_1 P-\lambda_2 R)}+\frac{2\lambda_2^2(1-R)}{(s+3\lambda_1+\lambda_C)} \end{array}\right] \tag{7.49}$$

$$\bar{P}_{down}(s) = 1-\bar{P}_{up}(s) \tag{7.50}$$

TABLE 7.2
Variation in $P_{up}(t)$ with Respect to Time

Time	Copula Repair	General Repair
0	1.00000	1.00000
10	0.93412	0.92019
20	0.84997	0.83859
30	0.77341	0.76421
40	0.70374	0.69644
50	0.64034	0.63467
60	0.58266	0.57839
70	0.53017	0.52709
80	0.48241	0.48034
90	0.43896	0.43774
100	0.39942	0.39892

7.6 ANALYTICAL STUDY

7.6.1 Availability Analysis

Let the repair facility follow both general and copula distribution, then we have $\bar{S}_{\mu_0}(s) = \bar{S}_{\exp\left[x^\theta + \{\log\phi(x)\}^\theta\right]^{1/\theta}}(s) = \dfrac{\exp\left[x^\theta + \{\log\phi(x)\}^\theta\right]^{1/\theta}}{s + \exp\left[x^\theta + \{\log\phi(x)\}^\theta\right]^{1/\theta}}$ and setting $\bar{S}_{\phi_i}(s) = \dfrac{\phi_i}{s+\phi_i}, i = 1, 2$. Choosing the values of various parameters as $\lambda_1 = 0.03$, $\lambda_2 = 0.035$, $\lambda_C = 0.025$, $\phi_i = 1$, $x = 1$, $\theta = 1 (i = 1, 2)$ in (7.49), and solving using Maple, we obtain the availability of the system as:

$$\begin{aligned} P_{up}(t) = {} & 0.011597e^{-2.750771t} - 0.000106e^{-1.247855t} + 0.001075e^{-1.152024t} \\ & + 0.054363e^{-1.113208t} + 1.026599e^{-0.009440t} - 0.017249e^{-1.115000t} \\ & - 0.026589e^{-1.120000t} - 0.005453e^{-1.125000t} \end{aligned} \tag{7.51}$$

Putting the various values of time variable, one may get different values of $P_{up}(t)$ with the help of (7.51) as shown in Table 7.2 and the corresponding Figure 7.3.

7.6.2 Reliability of the System

Take all repairs to zero and taking the failure rates as $\lambda_1 = 0.03$, $\lambda_2 = 0.035$ and $\lambda_C = 0.025$ in (7.49), and solving using Maple, we obtain the reliability of the system as:

$$\begin{aligned} R_i(t) = {} & 3.000000e^{-0.155000t} + 0.090000e^{-0.125000t} - 4.318846e^{-0.185000t} \\ & + 0.193846e^{-0.120000t} + 2.000000e^{-0.150000t} + 0.035000e^{-0.115000t} \end{aligned} \tag{7.52}$$

Putting the various values of time variable, one may get different values of reliability $R_i(t)$ with the help of (7.52) as shown in Table 7.3 and the corresponding Figure 7.4.

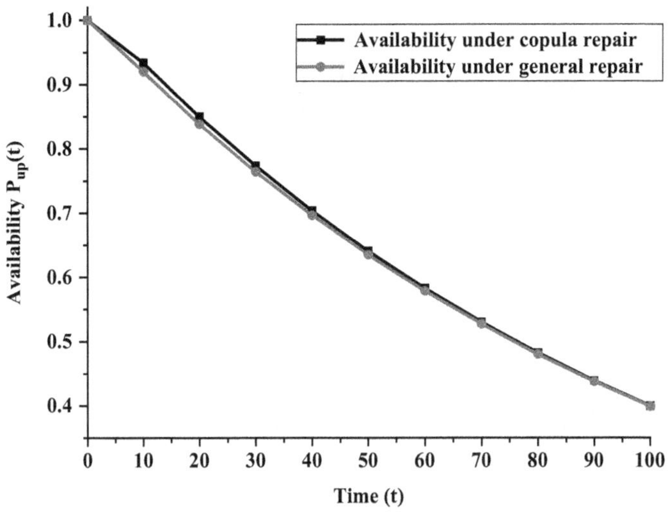

FIGURE 7.3 Availability as a function of time.

TABLE 7.3
Variation in $R_i(t)$ with Respect to Time

Time	$R_i(t)$
0	1.00000
10	0.49917
20	0.15642
30	0.04263
40	0.01095
50	0.00274
60	0.00068
70	0.00017
80	0.00004
90	0.00001
100	0.00000

7.6.3 Mean Time to Failure (MTTF)

Putting all repairs and s to zero in (7.49), we can receive the MTTF as:

$$MTTF = \frac{1}{\lambda}\left[\begin{array}{l} 1 + \dfrac{3\lambda_1}{2\lambda_1 + 2\lambda_2 + \lambda_C} + \dfrac{6\lambda_1^2}{\lambda_1 + 2\lambda_2 + \lambda_C} \\ + \dfrac{12\lambda_1\lambda_2}{2\lambda_1 + \lambda_2 + \lambda_C} + \dfrac{2\lambda_2}{3\lambda_1 + \lambda_2 + \lambda_C} + \dfrac{2\lambda_2^2}{3\lambda_1 + \lambda_C} \end{array}\right] \tag{7.53}$$

where $\lambda = 3\lambda_1 + 2\lambda_2 + \lambda_C$.

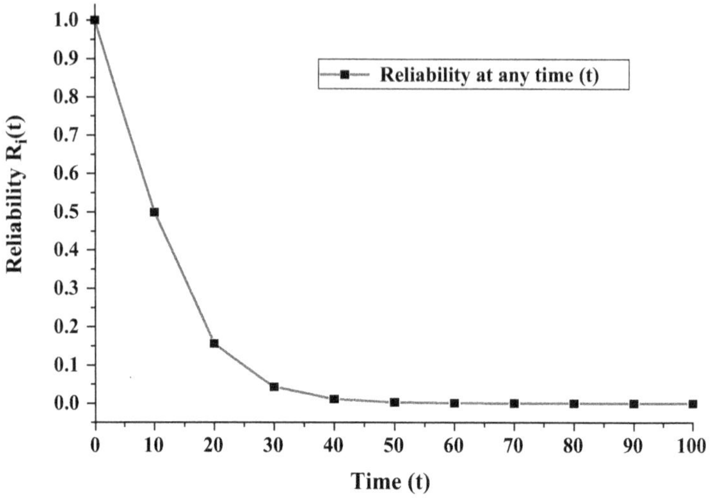

FIGURE 7.4 Reliability as a function of time.

TABLE 7.4
MTTF Variation Corresponding to λ_1, λ_2 and λ_C

Time (t)	λ_1	λ_2	λ_C
0.01	17.131	15.768	13.852
0.02	13.944	13.862	12.553
0.03	11.982	12.520	11.457
0.04	10.616	11.510	10.522
0.05	9.596	10.717	9.716
0.06	8.800	10.073	9.015
0.07	8.158	9.538	8.401
0.08	7.629	9.086	7.860
0.09	7.184	8.700	7.379
0.10	6.805	8.367	6.949

Taking failure rates as $\lambda_1 = 0.03$, $\lambda_2 = 0.035$, $\lambda_C = 0.025$and varying each λ_1, λ_2 and λ_C one by one in (7.53), the variation in MTTF can be found as given in Table 7.4 and Figure 7.5.

7.6.4 Cost Analysis

Let the service facility be available full-time, then expected profit may be incurred in the interval (0, t) using the following formula

$$E_p(t) = K_1 \int_0^t P_{up}(t)\,dt - K_2 t \tag{7.54}$$

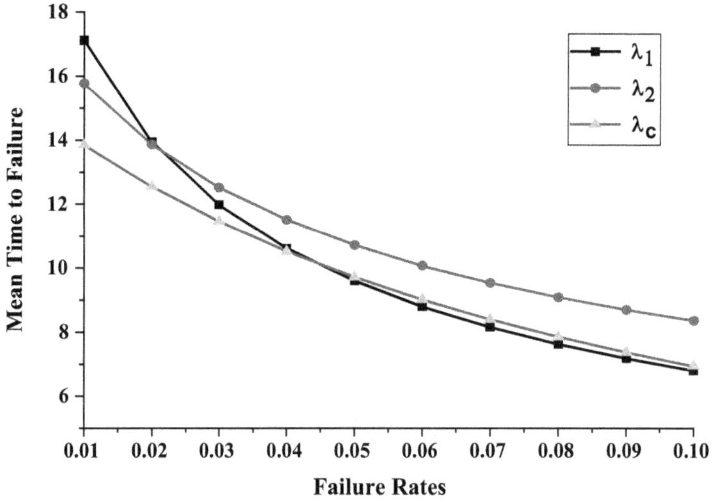

FIGURE 7.5 MTTF as a function of failure rates.

For the same set of parameters set in (7.49), one can obtain (7.55). Therefore,

$$
\begin{aligned}
E_p(t) = &-108.750957e^{-0.009439t} - 0.004216e^{-2.750771t} + 0.000085e^{-1.247855t} \\
&-0.048834e^{-1.113208t} - 0.000933e^{-1.152024t} + 0.015470e^{-1.115000t} \\
&-0.023740e^{-1.120000t} - 0.004846e^{-1.125000t} + 108.760799 - K_2 t
\end{aligned} \tag{7.55}
$$

Setting $K_1 = 1$and varying K_2 and time variable t, expected profit can be seen for copula repair in Table 7.5 and Figure 7.6 and for general repair Table 7.6 and Figure 7.7.

TABLE 7.5
Profit Calculation for Various t Under Copula Repair

	K_2					
Time (t)	**0.6**	**0.5**	**0.4**	**0.3**	**0.2**	**0.1**
0	0	0	0	0	0	0
10	3.81	4.81	5.81	6.81	7.81	8.81
20	6.72	8.72	10.72	12.72	14.72	16.72
30	8.83	11.83	14.83	17.83	20.83	23.83
40	10.21	14.21	18.21	22.81	26.81	30.81
50	10.93	15.93	20.93	25.93	30.93	35.93
60	11.04	17.04	23.04	29.04	35.04	41.04
70	10.60	17.60	24.60	31.60	38.60	45.60
80	9.66	17.66	25.66	33.66	41.66	49.66
90	8.26	17.26	26.26	35.26	44.26	53.26
100	6.49	16.49	26.49	36.49	46.49	56.49

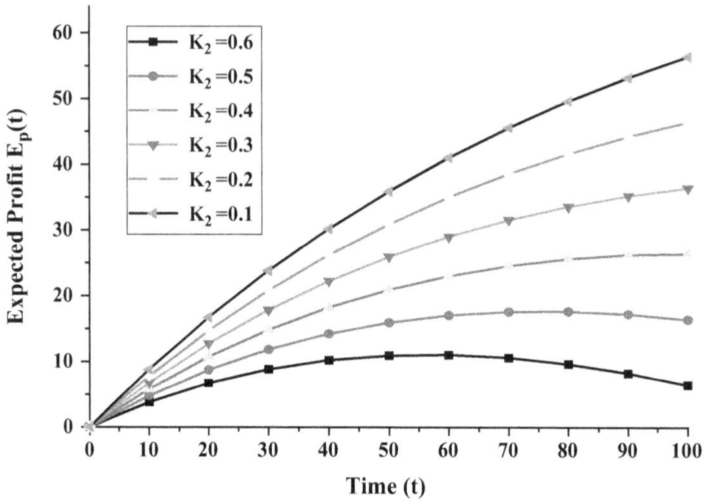

FIGURE 7.6 Expected profit as a function of time under copula repair.

TABLE 7.6
Profit Calculation for Various t Under General Repair

	K_2					
Time (t)	0.6	0.5	0.4	0.3	0.2	0.1
0	0	0	0	0	0	0
10	3.61	4.61	5.61	6.61	7.61	8.61
20	6.40	8.40	10.40	12.40	14.40	16.40
30	8.41	11.41	14.41	17.41	20.41	23.41
40	9.71	13.71	17.71	21.71	25.71	29.71
50	10.36	15.36	20.36	25.36	30.36	35.36
60	10.42	16.42	22.42	28.42	34.42	40.42
70	9.94	16.94	23.94	30.94	37.94	44.94
80	8.98	16.98	24.98	32.98	40.98	48.98
90	7.56	16.56	25.56	34.56	43.56	52.56
100	5.74	15.74	25.74	34.74	44.74	54.74

7.7 CONCLUSION

In this chapter, we measured the reliability characteristics of a 3-out-of-5: G warm standby system consisting of three priority units and two supporting units in parallel configuration with catastrophic failure. Warm-standby redundancy, including copula repair, has been used as an effective method to achieve a more reliable system design. Through computations in the analytical section, we have noticed that the system performs well when the copula repair is carried out. The following interpretations can be drawn on the basis of the analysis in this article:

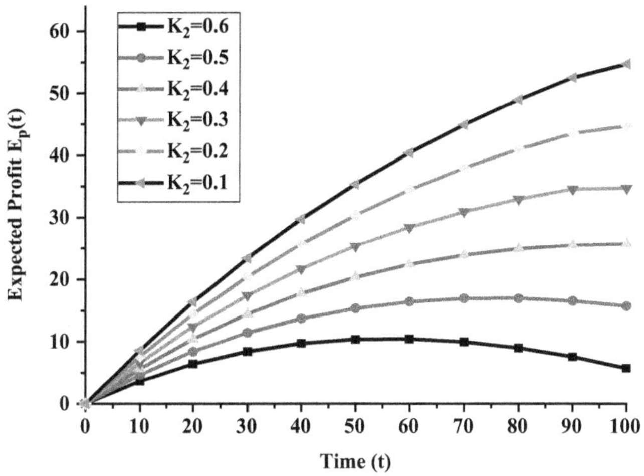

FIGURE 7.7 Expected profit as a function of time under general repair.

Table 7.2 and Figure 7.3 present the availability of the system in two situations under copula repair for complete failed states and general repair for partially failed states for the fixed values of failure rates. Figure 7.3 indicates that availability is little better when repair follows copula distribution. The availability of the repairable system decreases with respect to the time, when failure rates are set at $\lambda_1 = 0.03$, $\lambda_2 = 0.035$, $\lambda_C = 0.025$, and ultimately becomes steady to the value zero after a sufficient long interval of time. Thus, the future behavior of a complex system can be safely predicted at any time for any given set of parametric values, as can be seen from the graphical consideration of the model. Table 7.3 and Figure 7.4 demonstrate the change in reliability from one to zero with respect to time t under no repair. Obviously, it can be seen that the reliability values are lower than availability values for the same value of failure rates. Thus, it can be observed that routine repairs are required for better output of the repairable systems. Table 7.4 and Figure 7.5 reveals the MTTF of the system with respect to variation in failure in all the three cases. We can realize that the MTTF of the system decreases gradually as failure rates increases. MTTF found to be highest for λ_2.

When revenue cost per unit time K_1 is fixed at 1 and service costs $K_2 = 0.6, 0.5, 0.4, 0.3, 0.2, 0.1$, profit is estimated and results can be seen in Figure 7.6 when the repair follows Gumbel–Hougaard family copula distribution and Figure 7.7, when repair follow general distribution. A critical review of Figures 7.6 and 7.7 reveals that the estimated profit varies with respect to the time, but projected profit attributed to copula repair is higher than general repair. This is a clear indicator that copula repair improves availability and therefore increases profit. At the end of the day, one can observe that as the service cost increases, profit declines. In general, the expected profit for low service costs is high relative to high service costs.

The model developed in this chapter has been found to be very useful for power generation and transmission systems, as we have five equipment groups and are helpful in proper maintenance analysis, decision-making and performance assessment.

We may expand our work to a number of methods, such as the Kolmogorov method and the fuzzy reliability method, by considering repair rates as constant. Furthermore, the system can be analyzed by taking k-out-of-(m+n): G/F policy.

REFERENCES

1. Amari, S.V., Pham, H. and Misra, R.B. (2012), "Reliability characteristics of k-out-of-n warm standby systems", *IEEE Transactions on Reliability*, Vol. 61, pp. 1007–1018.
2. Gera, A.E. (2004), "Combined k-out-of-n:G, and consecutive k/sub c/-out-of-n:G systems", *IEEE Transactions on Reliability*, Vol. 53(4), pp. 523–531.
3. Huang, J., Zuo, M.J. and Wu, Y. (2000), "Generalized multi-state k-out-of-n:G systems", *IEEE Transactions on Reliability*, Vol. 49(1), pp. 105–111.
4. Levitin, G. and Amari, S.V. (2010), "Approximation algorithm for evaluating time-to-failure distribution of k-out-of-n system with shared standby elements", *Reliability Engineering and System Safety*, Vol. 95, pp. 396–401.
5. Liang, X., Xiong, Y. and Li, Z. (2010), "Exact reliability formula for consecutive k-out-of-n repairable systems", *IEEE Transactions on Reliability*, Vol. 59(2), pp. 313–318.
6. Myers, A.F. (2007), "k-out-of-r-from-n: G system reliability with imperfect fault coverage", *IEEE Transactions on Reliability*, Vol. 56(3), pp. 464–473.
7. Park, M. and Pham, H. (2012), "A generalized block replacement policy for a k-out-of-n system with respect to a threshold number of failed components and risk costs", *IEEE Transactions on Systems, Man, and Cybernetics - Part A: Systems and Humans*, Vol. 42(2), pp. 453–463.
8. Salehi, E.T., Asadi, M. and Eryilmaz, S. (2011), "Reliability analysis of consecutive k-out-of-n systems with non-identical components lifetimes", *The Journal of Statistical Planning and Inference*, Vol. 141(8), pp. 2920–2932.
9. She, J. and Pecht, M.G. (1992), "Reliability of a k-out-of-n warm-standby system", *IEEE Transactions on Reliability*, Vol. 41, pp. 72–75.
10. Xing, L., Shrestha, A. and Dai, Y. (2011), "Exact combinatorial reliability analysis of dynamic systems with sequence-dependent failures", *Reliability Engineering and System Safety*, Vol. 96(10), pp. 1375–1385.
11. Yaghoubi, A., Naiki, S.T.A. and Rostamzadeh, H. (2020), "A closed-form equation for steady-state availability of cold standby repairable k-out-of-n: G system", *International Journal of Quality Reliability and Management*, Vol. 37(1), pp. 145–154.
12. Kumar, P. and Gupta, R. (2007), "Reliability analysis of a single unit M|G|1 system model with helping unit", *Journal of Combinatorics, Information & System Sciences*, Vol. 32(1–4), pp. 209–219.
13. Ram, M., Singh, S. B. and Singh, V. V. (2013), "Stochastic analysis of a standby system with waiting repair strategy", *IEEE Transactions on System, Man, and Cybernetics System*, Vol. 43(3), pp. 698–707.
14. Singh, V.V. and Poonia, P.K. (2019), "Probabilistic assessment of two units parallel system with correlated lifetime under inspection using regenerative point technique", *International Journal of Reliability, Risk and Safety: Theory and Application*, Vol. 2(1), pp. 5–14.
15. Tiwari, N. and Singh, S.B. (2014), "Reliability and sensitivity analysis of a two unit warm standby system with low efficiency unit", *Elixir Statistics*, Vol. 69, pp. 23185–23203.
16. Yusuf, I. (2016), "Reliability modelling of a parallel system with a supporting device and two types of preventive maintenance", *International Journal of Operational Research*, Vol. 23(3), pp. 269–287.

17. Srinivasan, S. K. and Subramanian, R. (2006), "Reliability analysis of a three unit warm standby redundant system with repair", *Annals of Operation Research*, Vol. 143, pp. 227–235.
18. Munjal, A. and Singh, S. B. (2014), "Reliability analysis of a complex repairable system composed of two 2-out-of-3: G subsystems connected in parallel", *Journal of Reliability and Statistical Studies*, Vol. 7, pp. 89–111.
19. Goyal, N., Ram, M., Amoli, S. and Suyal, A. (2017), "Sensitivity analysis of a three-unit series system under k-out-of-n redundancy", *International Journal of Quality & Reliability Management*, Vol. 34(6), pp. 770–784.
20. Gupta, R., Chaudhary, A. and Jaiswal. S. (2015), "Stochastic analysis of a three-unit complex system with active and passive redundancies and correlated failures and repair times", *Journal of Mathematical and Computational Science*, Vol. 5(5), pp. 694–707.
21. Nelsen, R.B. (2006), *An Introduction to Copulas*, 2nd edition. Springer, New York.
22. Singh, V.V., Poonia, P.K. and Rawal, D.K. (2020), "Reliability analysis of repairable network system of three computer labs connected with a server under 2-out-of-3: G configuration", *Life Cycle Reliability and Safety Engineering*, Vol. 10(1), pp.19–29. doi: 10.1007/s41872-020-00129-w.
23. Gehlot, M., Singh, V.V., Ayagi, H. I. and Goel, C.K. (2018), "Performance assessment of repairable system in series configuration under different types of failure and repair policies using Copula Linguistics", *International Journal of Reliability and Safety*, Vol. 12(4), pp. 367–374.
24. Lado, A., Singh, V.V., Ismail, K.H. and Ibrahim, Y. (2018), "Performance and cost assessment of repairable complex system with two subsystems connected in series configuration", *International Journal of Reliability and Applications*, Vol. 19(1), pp. 27–42.
25. Raghav, D., Poonia, P. K., Gahlot, M., Singh, V.V., Ayagi, H.I. and Adbullahi, A.H. (2020), "Probabilistic analysis of a system consisting of two subsystems in the series configuration under copula repair approach", *Journal of the Korean Society of Mathematical Education Series B-pure and Applied Mathematics*, Vol. 27(3), pp. 137–155.
26. Singh, V.V., Poonia, P.K. and Abdullahi, A.H. (2020), "Performance analysis of a complex repairable system with two subsystems in series configuration with an imperfect switch", *Journal of Mathematical and Computational Science*, Vol. 10(2), pp.359–383.
27. Poonia, P.K. and Sirohi, A. (2020), "Cost benefit analysis of a k-out-of-n: G type warm standby series system under catastrophic failure using copula linguistics", *International Journal of Reliability, Risk and Safety: Theory and Application*, Vol. 3(1), pp. 35–44.
28. Poonia, P. K., Sirohi, A. and Kumar, A. (2020), "Cost analysis of a repairable warm standby k-out-of-n: G and 2-out-of-4: G series systems under catastrophic failure using copula repair", *Life Cycle Reliability and Safety Engineering*, Vol. 10(2), pp. 121–133. doi: 10.1007/s41872-020-00155-8.

8 An Inventory Policy for Increasing Holding Cost under the Effect of Stock-Dependent Deterioration and Partial Backlogging

Abhinav Goel and Mohd Aftab Ali

CONTENTS

8.1 INTRODUCTION

In our routine life, the practical experience tells that some but not all clients can wait for the backlogged items for the period of shortages for stylish merchandises as well as technologically advanced products with the least living life cycle. If increase the waiting time due to any kind of circumstances, the backlogging price will be decreased Accordingly to this phenomenon, taking the backlogging price as an essential tool. Conversely, maximum of the inventory models is based on unrealistic situations; when there is no stock then all demand is either lost or

DOI: 10.1201/9781003089636-8

backlogged. In the real-world situation, some customers, due to goodwill or personal relation or by any means, are often willing to wait until the merchandise comes in stock, especially if the waiting time is shortest, whilst some others are irritated and go to a different place. The merchandise backlogging price depends on the replenishment time. In consideration of food products, dairy goods, bread, green vegetables, fruits packed merchandise, etc. decreases their effectiveness with increases of time. Storage space conveniences are required to remain usable, such as types of merchandise from the deterioration. To perform longer or greater storage periods, it requires some supplementary specialized machines and apparatus. Accordingly, holding costs cannot be unchanged of the total period of storage and it increases when time increases. Inventory models considering increasing holding cost with time have been developed by several researchers; Zangwill (1966) urbanized the products for the multi-time duration's arrangement model for the backlogging phenomena, and model for the combination of inventory backorders with lost sales was discussed by Montgomery et al. (1973). An economic production quantity model of the deteriorating objects was developed via Wee (1993) in consideration of partially backordering. Subsequently, Abad (1996) developed the model used for dynamic pricing in favor of perishable merchandises. The Inventory model for the deteriorating merchandises was considered for the number of items discount and partially back-ordering objects via Wee (1999). Shortages for the inventory models is taken into consideration with partially backlogging items. Papachristos and Skouri (2000) gives the Economic Ordered quantity model for the limited preparation horizon under the unvarying type deterioration charge with backlogging dependent on time and demand is also taken variable. Thereafter, an inventory model considering deteriorating merchandise through variable type demand was discussed in a paper via Skouri or Papachristos (2003a); in the model shortage was also allowable and assumed to be partially backlogged. Ouyang et al. (2005) projected the economic order quantity models of the deteriorating merchandise with declining demand. In this model, they consider the variable backlogging rate and it depends on the waiting period of the coming replenishment cycle. The model using deteriorating merchandises and partial backlogging is also proposed in that model via Skouri and Papachristos (2003b). In this paper, the backlogging price is considered to be a function for the offing time the beginning of the next merchandise cycle which is the function of time of the moment this when demand occurs, and the succeeding product cycle starts. Dash et al. (2004) also considered the phenomena of holding cost, with shortages and variable demand rate. Chang et al. (2006a) developed an EOQ model for perishable items under stock-dependent selling rate and time-dependent partial backlogging. An inventory model for EOQ to determine optimal selling price the replacement number and its to-do list for partially backlogged merchandise was given by Chang et al. (2006b) later on Chang and Lin (2010) developed a model for stock dependent consumption rate with inflation for partially backlogged items. Elsayed and Teresi (2007) developed the optimal order level of the deteriorating items. Then Chung and Huang (2007) discussed the two warehouses inventory models for the deteriorating items. Lee and Hsu (2009) developed a two-warehouse model with time-varying demand price. Some researchers consider a warehouse with unlimited capacity because in the case of an owned and rented warehouse it is not always easy to select which type of warehouse should be filled first or vacated first. To overcome this type of problem, we consider a warehouse with unlimited capacity. Soni and Shah (2008) designed the mathematical model considering stock-varying demand with optimal ordering policy below the progressive fee plan surroundings. Subsequently, Singh and Malik (2009) proposed the two warehouses inventory model by means of inflation-inducing demand. Chang et al. (2010) discussing the improved model with optimal replenishment policy under the stock-based demand price. Khanra et al. (2011) developed inventory models for the decaying item by means of quadratic demand under the permissible delay payments condition. Goel and Pandey (2011) investigated a production policy for the inventory system with ramp-type demand with shortages under inflation. Goel and Pandey (2012) investigated the flexible developed inventory model with permissible delay in payment as well as cash discount under the effects of time value money and inflation. In another article, Singh et al. (2013) produced an EOQ model with preservative technology investment when demand depends on selling price and credit period under two levels of trade credit. Vashistha et al. (2015) urbanized the optimum inventory system with maximum lifetime items; in their model, demand price is assumed to be a function of price and is also stockpile-dependent. Kumar et al. (2017) demonstrate an optimal inventory model assuming within-variable holding and sales revenue costs. M. Shahabi et al. (2018) developed a combined production inventory location problem with multi-variable demands. T. Sekar and R. Uthaya Kumar (2018) developed the production inventory models for only one vendor and only one buyer incorporated demand with various productions setup & rework. Ruidas et al. (2019) developed Single-phase production Inventory Model for Price reconsideration while Liu et al. (2019) recommended a service system involving the replacement of the inventory for the multi-client items with decreased replacement costs. Harit et al. (2020) develop a model considering the effect of preservation technology on the optimized use of two warehouses for deteriorating items.

8.2 ASSUMPTIONS AND NOTATIONS

The following notations and assumptions have been adopted in order to develop the proposed model.

The model is developed for only one deteriorating product with linearly price- and stock-dependent-type demand pattern,

$$D(p)=\begin{cases} a-bp+\alpha q(t) & \text{when } q(t)>0 \\ a-bp & \text{when } q(t)\leq 0 \end{cases}$$

where $a - bp > 0$

(1) The deterioration rate $\theta(0 < \theta << 1)$ is taken as constant and is depends on the stock.

(2) There is no available replacement and repair for the deteriorated merchandises for the period under consideration
(3) When the demand for goods is more than the supply, shortage will occur. A customer encountering shortage will either wait for the vender to reorder (backlogging cost involved) or go to other vendors. In this chapter shortages are allowed under the backlogging rate$[1 + \delta(T - t)]^{-1}$, where $(T - t)$ is the waiting time backlogging parameter δ is positive constant
(4) The holding cost per unit cycle is comparative for the time period of the storage of every all unit and is to be comparative to the unit purchasing cost c.
(5) Unit of purchasing cost $c$in the discount environment is a not a continuous and decreasing step function which depends on the number of order quantity (Q).

NOMENCLATURE

Notations	Description
A	The replenishment cost/per order cycle.
a	The constant part for demand rate (a>0).
B	The coefficient for the price in the demand rate (b>0).
C	The purchasing cost/per unit.
P	The selling price/per unit.
S	The shortage cost/per unit.
L	The opportunity cost/per unit.
θ	The deteriorating rate. $(0 \leq \theta \leq 1)$
G	The constant part for holding cost, fraction of unit purchase cost
H	Coefficient for linearly time varying holding cost, as fraction of unit purchase cost.
q (t)	The inventory level.
α	Stock-dependent demand rate parameter $(\alpha > 0)$
δ	Backlogging parameter
B	The initial number of inventories (for model with shortages).
R	The maximum number of partially backlogging quantity (for model with shortages).
Q	The number of order quantity per replenishment cycle.
R	Rate of inflation
t_1	Time at which the stock reaches zero (for model with shortages)
T	Length of each replenishment cycle (for both models)
$TC_1(T)$	Total cost for model without shortages
$TC_2(T)$	Total cost for model with shortages

8.3 MODEL FORMULATION

In this study we consider a situation where a person associated with the retail purchasing of deteriorating goods as of its provider in a discounted environment with a purchasing cost/per unit piece decreasing base going on the number of the order quantity. First the inventory model with no shortage in addition then an inventory types of model with partial backlogging. Shortages will also be discussing underneath the circumstance of inflation when permissible delay in the payments is also allowed.

8.4 MODEL WITHOUT SHORTAGES

In these cases, an inventory level is Q at time t = 0. Owing to the combined special effects of demands the deterioration an inventory level will be zero at that time $t = T$later than some time replenishment will be made for occupied inventory system and do again the entire inventory system (see Figure 8.1). Therefore, the level of inventory at a few instantaneous can be defined through the subsequent governing differential equation.

$$\frac{dq(t)}{dt} + \theta q(t) = -(a - bp + \alpha q(t)), \qquad 0 \le t \le T \tag{8.1}$$

With the boundary conditions $q(0) = Q$ and $q(T) = 0$

Equation (8.1) can be written as follow

$$\frac{dq(t)}{dt} + \eta q(t) = -D, \qquad 0 \le t \le T, \tag{8.2}$$

Wherever we consider $D = a - bp,\ \eta = \alpha + \theta$.

Solution to Equation (8.2) refers for an inventory level for any given time.

$$q(t) = \frac{D}{\eta}\left(e^{\eta(T-t)} - 1\right) \tag{8.3}$$

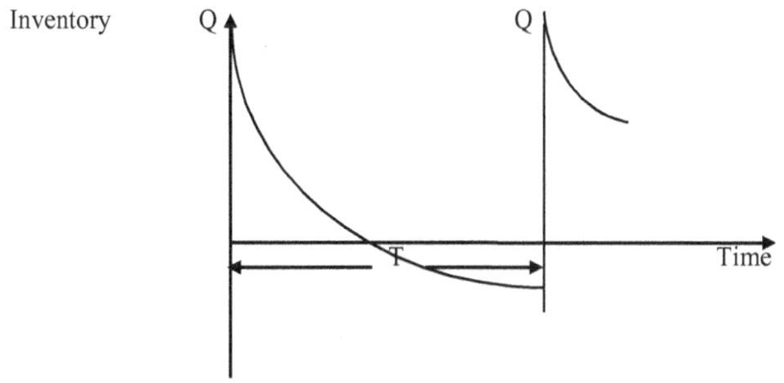

FIGURE 8.1 Graphical representation of proposed inventory model with no shortages.

One can derive using the initial condition from Equation (8.3)

$$Q = \frac{D}{\eta}\left(e^{\eta T} - 1\right). \tag{8.4}$$

So the length of entire cycle can be expressed as follow.

$$T = \frac{1}{\eta}\left(\ln 1 + \frac{\eta Q}{D}\right). \tag{8.5}$$

Total cost for this proposed inventory system has the following mechanism:

Ordering cost (OC): A

Holding cost (HC):

$$c\int_0^T (g + ht)q(t)e^{-rt}dt$$

$$= \frac{c(a-bp)}{\eta}\left(\begin{array}{l} g\left(\dfrac{e^{\eta T}}{(\eta + r)} + \dfrac{\eta e^{-rT}}{r(\eta + r)} - \dfrac{1}{r}\right) \\ +h\left(\dfrac{\eta e^{-rT}(\eta + 2r)}{r^2(\eta + r)^2} + \dfrac{Te^{-rT}}{r(\eta + r)} + \dfrac{e^{\eta T}}{(\eta + r)^2} - \dfrac{1}{r^2}\right)\end{array}\right)$$

Purchasing cost (PC): $\dfrac{c(a-bp)}{\eta}\left(e^{\eta T} - 1\right).$

Hence total cost for the inventory

$$TC_1 = \frac{1}{T}\left(OC + HC + PC + IP - IE\right)$$

$$TC_1 = \frac{1}{T}\left[\begin{array}{l} A + \dfrac{c(a-bp)}{\eta}\left(\begin{array}{l} g\left(\dfrac{e^{\eta T}}{(\eta + r)} + \dfrac{\eta e^{-rT}}{r(\eta + r)} - \dfrac{1}{r}\right) \\ +h\left(\dfrac{\eta e^{-rT}(\eta + 2r)}{r^2(\eta + r)^2} + \dfrac{Te^{-rT}}{r(\eta + r)} + \dfrac{e^{\eta T}}{(\eta + r)^2} - \dfrac{1}{r^2}\right)\end{array}\right) \\ + \dfrac{c(a-bp)}{\eta}\left(e^{\eta T} - 1\right)\end{array}\right] \tag{8.6}$$

8.5 MODEL WITH PARTIAL SHORTAGES

On commencement of the cycle order of $Q = (R + S)$units this piece comes into stockpile. Then the units are utilizing to meet a total accumulate backlogging demand

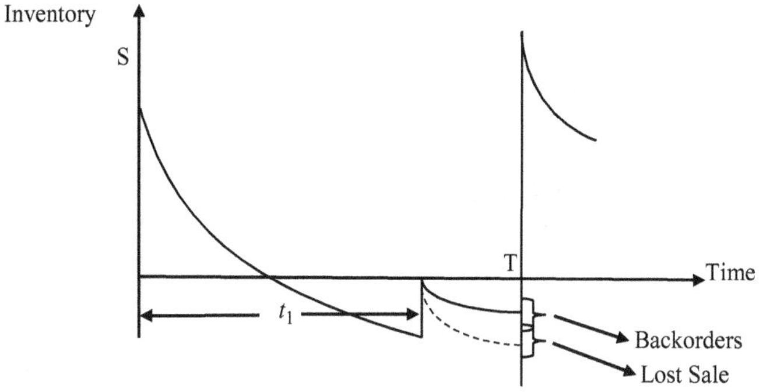

FIGURE 8.2 Graphical representation of proposed an inventory model with shortages.

resultant in an on-stock level 'S'. The merchandises will be consumed to meet customer demand and spoil at the constant rate at some stage in the time interval. Owing to customer demand as well as resulting impact of decline the inventory level of the decreases to zero t = t_1. Shortly afterwards, the shortages appear, which accumulates during period $[t_1, T]$and depends on wait for time of a clients. At this time the new-fangled order is re-created and the entire inventory system is replicated. The performance of the entire inventory system at full cyclic length can be shown in the Figure 8.2. Then the position of an inventory level q(t) at any instantaneous $t \in [0, T]$ can be present via the subsequent governing differential equation

$$\frac{dq_1(t)}{dt} + \theta q_1(t) = -\left[a - bp + \alpha q_1(t)\right], \qquad 0 < t \le t_1 \tag{8.7}$$

And

$$\frac{dq_2(t)}{dt} = \frac{-(a-bp)}{1+\delta(T-t)}, \qquad t_1 < t \le T \tag{8.8}$$

With the boundary conditions $q_1(0) = S$, $q_1(t_1) = 0$ and $q_2(T) = -R$.

Equation (8.7) can be rewritten as follows.

$$\frac{dq_1(t)}{dt} + \eta q_1(t) = -D, \qquad 0 < t \le t_1 \tag{8.9}$$

Where $D = a - bp$, $\eta = \alpha + \theta$.

The solution of Equation (8.9) using $q_1(t_1) = 0$ and $q_2(T) = -R$ is given by

$$q_1(t) = \frac{D}{\eta}\left(e^{\eta(t_1 - t)} - 1\right) \tag{8.10}$$

And

$$q_2(t) = \frac{D}{\delta}\ln\left(1+\delta\left(T-t\right)\right) - R. \tag{8.11}$$

Utilizing the initial condition $q_1(0) = S$ in Equation (8.10) one can get the initial stock.

$$S = \frac{D}{\eta}\left(e^{\eta t_1} - 1\right). \tag{8.12}$$

Once more with the help of continuity of $q(t)$ at $t = t_1$ the maximum shortage level is given below.

$$R = \frac{D}{\delta}\ln\left(1+\delta\left(T-t_1\right)\right). \tag{8.13}$$

Therefore, the total number of ordering quantity at the beginning of the cycle is given below.

$$Q = \frac{D}{\eta}\left(e^{\eta t_1} - 1\right) + \frac{D}{\delta}\ln\left(1+\delta\left(T-t_1\right)\right). \tag{8.14}$$

The profit function of the inventory system has the subsequent mechanism.

Ordering cost (OC): A.

Holding cost (HC):

$$c\int_0^{t_1}\left(g+ht\right)q_1\left(t\right)e^{-rt}dt$$

$$= \frac{c\left(a-bp\right)}{\eta}\left(\begin{array}{l} g\left(\dfrac{e^{\eta t_1}}{\left(\eta+r\right)} + \dfrac{\eta e^{-rt_1}}{r\left(\eta+r\right)} - \dfrac{1}{r}\right) \\ +h\left(\dfrac{\eta e^{-rt_1}\left(\eta+2r\right)}{r^2\left(\eta+r\right)^2} + \dfrac{t_1\eta e^{-rt_1}}{r\left(\eta+r\right)} + \dfrac{e^{\eta t_1}}{\left(\eta+r\right)^2} - \dfrac{1}{r^2}\right)\end{array}\right)$$

Purchasing cost (PC):

$$= c\left(a-bp\right)\left[\frac{1}{\eta}\left(e^{\eta t_1} - 1\right) + \frac{1}{\delta}\ln\left(1+\delta\left(T-t_1\right)\right)\right]$$

Shortage cost (SC):

$$= -s\int_{t_1}^{T} q_2(t)e^{-rt}dt$$

$$= \frac{-s(a-bp)}{r}\left[e^{-rt_1}(T-t_1)+\frac{((a-bp)+Rr)}{(a-bp)r}\left(e^{-rT}-e^{-rt_1}\right)\right]$$

Lost Sale cost (LSC):

$$= l\int_{t_1}^{T}\left(1-\frac{1}{(1+\delta(T-t))}\right)(a-bp)e^{-rt}dt$$

$$= \frac{l(a-bp)\delta}{r}\left(e^{-rt_1}\left(T-t_1+\frac{1}{r}\right)-\frac{1}{r}e^{-rT}\right)$$

$$TC_2 = \frac{1}{T}\left[\begin{array}{l} A+\frac{c(a-bp)}{\eta}\left(\begin{array}{l} g\left(\frac{e^{\eta t_1}}{(\eta+r)}+\frac{\eta e^{-rt_1}}{r(\eta+r)}-\frac{1}{r}\right)+ \\ h\left(\frac{\eta e^{-rt_1}(\eta+2r)}{r^2(\eta+r)^2}+\frac{t_1\eta e^{-rt_1}}{r(\eta+r)}+\frac{e^{\eta t_1}}{(\eta+r)^2}-\frac{1}{r^2}\right) \end{array}\right)+ \\ c(a-bp)\left[\frac{1}{\eta}\left(e^{\eta t_1}-1\right)+\frac{1}{\delta}\ln\left(1+\delta(T-t_1)\right)\right]- \\ \frac{s(a-bp)}{r}\left[e^{-rt_1}(T-t_1)+\frac{((a-bp)+Rr)}{(a-bp)r}(e^{-rT}-e^{-rt_1})\right]+ \\ \frac{l(a-bp)\delta}{r}\left(e^{-rt_1}\left(T-t_1+\frac{1}{r}\right)-\frac{1}{r}e^{-rT}\right) \end{array}\right]$$

Currently, the merchant objective to find out the optimal period of positive inventory level.

t_1^* Replenishment T^* in the order to be minimizes the merchant the total cost/ per unit time. The retailer here aims to find out the maximum period of the positive the level of inventory as well as replenish to reduce the seller total cost/per unit time.

The total cost function is $TC_1(T) = \frac{f_1(T)}{g_1(T)}$.

8.6 OPTIMAL CRITERIA

The first- and second-order derivatives of $f_1(T)$ with respect to T are

$$f_1'(T)=c(a-bp)\left(\frac{g}{(\eta+r)}\left(e^{\eta T}-e^{-rT}\right)+h\left(\frac{e^{\eta T}}{(\eta+r)^2}-\frac{(\eta+2r)e^{-rT}}{r(\eta+r)^2}+\frac{e^{-rT}}{\eta r(\eta+r)}(1-Tr)\right)+e^{\eta T}\right)$$

$$f_1''(T)=c(a-bp)\left(\frac{g}{(\eta+r)}\left(\eta e^{\eta T}+re^{-rT}\right)+h\left(\frac{\eta e^{\eta T}}{(\eta+r)^2}+\frac{(\eta+2r)e^{-rT}}{(\eta+r)^2}+\frac{e^{-rT}}{\eta(\eta+r)}(Tr-2)\right)+\eta e^{\eta T}\right)>0$$

$$\text{where } Tr>2$$

Therefore the $f_1(T)$ is a non-negative and differentiable and convex function. Also, with respect to T$g_1(T)=T$ is positive, differentiable and concave function. $g_2(t_1,T)=T>0$.

We require to show that the $f_2(t_1,T)$ is non-negative and differentiable and (strictly) joint convex function with respect to the t_1 and T. In the direction of construct the Hessian matrix for $f_2(t_1,T)$ then the calculated at all the second-order partial derivatives with respect to t_1 and T.

$$\frac{\partial^2 f_2(.)}{\partial t_1^2}=(a-bp)\left(\begin{array}{l}\frac{e^{-M(r+\eta)-rt_1}\left(re^{M(r+\eta)}+e^{(r+\eta)t_1}\eta\right)i_pc}{(r+\eta)}+\\ e^{-rt_1}\delta l(3+rT-rt_1)+e^{-rt_1}s(-2-rT+rt_1)\\ +\frac{e^{-rt_1}c\left(\begin{array}{l}\left(-1+e^{(r+\eta)t_1}\right)h\eta+g(r+\eta)\\ \left(r+e^{(r+\eta)t_1}\eta\right)+hr(r+\eta)t_1\end{array}\right)}{(r+\eta)^2}+\\ c\left(\eta e^{\eta t_1}-\frac{\delta}{(1+T\delta-\delta t_1)^2}\right)\end{array}\right)$$

$$\frac{\partial^2 f_{2.1}(.)}{\partial T\partial t_1}=\frac{e^{-r(M-t_1)}(a-bp)\left(\begin{array}{l}e^{-rt_1}\left(e^{-Mr}-1\right)\delta pi_e-\\ e^{Mr}r\left(e^{rt_1}\delta c-(\delta l-s)(1+T\delta-\delta t_1)^2\right)\end{array}\right)}{r(1+T\delta-\delta t_1)^2}$$

Therefore, the Hessian matrix for $f_2(t_1, T)$

$$H_{ii} = \begin{bmatrix} \frac{\partial^2 f_2(.)}{\partial t_1^2} & \frac{\partial^2 f_2(.)}{\partial t_1 \partial T} \\ \frac{\partial^2 f_2(.)}{\partial T \partial t_1} & \frac{\partial^2 f_2(.)}{\partial T^2} \end{bmatrix}.$$

Here both the principal minor $|H_{11}| = \frac{\partial^2 f_2(.)}{\partial t_1^2} > 0$ and $|H_{22}| = \frac{\partial^2 f_2(.)}{\partial t_1^2}\frac{\partial^2 f_2(.)}{\partial T^2} - \frac{\partial^2 f_2(.)}{\partial T \partial t_1}\frac{\partial^2 f_2(.)}{\partial t_1 \partial T} > 0$

Because each one the principal minors of Hessian matrix for the $f_2(t_1, T)$ are positive and Hessian matrix is positive exact.

8.7 NUMERICAL ILLUSTRATION

In order to optimize the solution of the presented study, results and to achieve managerial insights of our project, here we illustrate mathematical examples for two different cases.

8.7.1 Illustration for Model Without Shortages

For the validation of the proposed model without shortages, consider the values of the following parameters A = \$150/Order, a = 350, b = 3, g = \$2.1/unit, h = \$2.3/unit, p = \$3/unit, $\theta = 0.01$, r = 0.6, c = \$0.5/unit, $\eta = 0$.

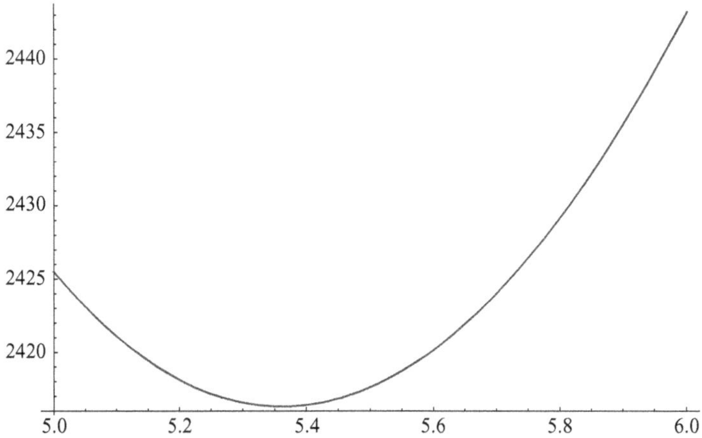

The Optimal solution of the model without shortages by using above optimal criteria we obtained $T \rightarrow 5.3$ and $TC_{1.1} \rightarrow \$2400.32$ which is shown in above figure.

Example. Model with Shortages

Data are similar those are use in illustration 1. Additionally we consider that the $\delta = .78$, $C_s = 40.5$ and $C_l = 32.6$, for the backlogged shortages.

Solution:
Optimal Total cost =$669.931, T=2.803 year, t_1= 1.475 year

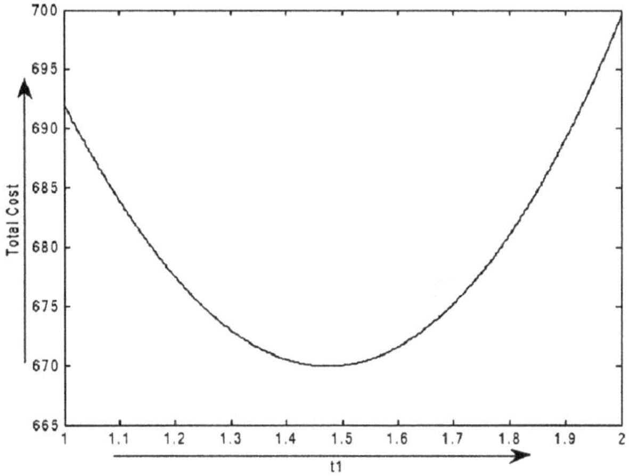

Total cost versus t_1.

Total cost versust_1.

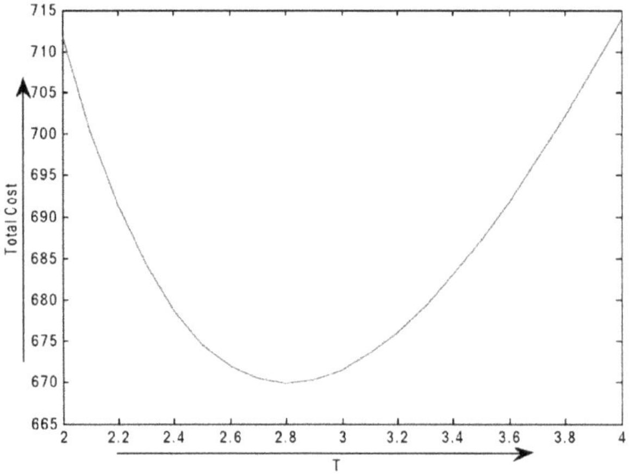

Total cost versus T.

Total cost versus T.

8.8 SENSITIVITY ANALYSIS

Consider the effect of relevant parameters for zero ending case

8.9 OBSERVATIONS

A sensitivity analysis is performed here and on its basis these following are the Observations.

- Table 8.1 lists the variation in the holding cost parameter 'A'. It is observed from the table that with increment holding cost A value of critical time 'T' increases and value of the total Cost (TC_1) increases.
- Table 8.2 lists the variations in coefficient of price in demand rate parameter 'b' increases value of critical time 'T' decrease and the value of Total Cost (TC_1) decreasing.

TABLE 8.1
Sensitivity Analysis for Variations in Holding Cost (A)

A	Total Cost (TC_1)	Time (T)
+20%	5489	2.97
+15%	5373	2.49
+10%	4865	2.98
+5%	4561	2.98
0%	3425	2.57
–5%	3210	2.12
–10%	–2910	1.95
–15%	–2431	1.32
–20%	–1983	1.5

TABLE 8.2
Sensitivity Analysis for Variations in Coefficient of Price in Demand Rate (b)

B	Total Cost (TC_1)	Time (T)
+20%	1981	3.72
+15%	2310	3.12
+10%	2218	3.01
+5%	3210	2.98
0%	3425	2.57
–5%	3981	1.98
–10%	4120	1.59
–15%	4231	1.39
–20%	4531	1.38

- Table 8.3 lists the variations in constant part of holding cost parameter 'g' increases value of critical time 'T' increases and value of the total Cost (TC_1) increases.
- Table 8.4 lists for the time varying holding parameter 'h' increases the value of critical time 'T' increases and value of the total Cost (TC_1) increases.
- Table 8.5 lists for the variations in inflation rate parameter 'r' increases the value of critical time 'T' increases and value of the total Cost (TC_1) increases.
- Table 8.6 lists the variations in interest paid rate parameter (I_p). increases the value of critical time 'T' decrease and value of the total Cost (TC_1) increase and decrease.

TABLE 8.3
Sensitivity Analysis for Time Varying Holding Cost (h)

h	Total Cost (TC_1)	Time (T)
+20%	4401.25	3.608
+15%	4170.14	3.394
+10%	3931.47	3.154
+5%	3683.78	2.884
0%	3425	2.57
−5%	3415	2.28
−10%	3324	2.18
−15%	3211.32	2.01
−20%	3153.21	1.32

TABLE 8.4
Sensitivity Analysis for Variations in Constant Part of Holding Cost (g)

G	Total Cost (TC_1)	Time (T)
+20%	3965.98	2.023
+15%	3523.89	2.188
+10%	3478.62	2.31
+5%	3455.78	2.477
0%	3425	2.57
−5%	3393.22	2.681
−10%	3359.54	2.786
−15%	−3325.29	2.894
−20%	−3289.21	3.004

TABLE 8.5
Sensitivity Analysis for Variations in Inflation Rate (r)

r	Total Cost (TC_1)	Time (T)
+20%	5898.10	2.37
+15%	5672.98	2.48
+10%	4725.43	2.53
+5%	4610.9	2.76
0%	3425	2.57
−5%	3004.73	2.873
−10%	2758.06	3.125
−15%	2267.7	3.446
−20%	2174.64	3.806

TABLE 8.6
Sensitivity Analysis for Variations in Interest Paid Rate (I_p)

I_p	Total Cost (TC_1)	Time (T)
+20%	3321.12	2.12
+15%	3451.12	2.25
+ 10%	3512.87	2.28
+5%	3489.09	2.37
0%	3425	2.57
−5%	3391.11	2.775
−10%	−3382.12	2.798

- Table 8.7 shows the variations in interest earned rate (I_e). I increase the value of critical time 'T' decrease and the value of Total Cost (TC_1) decrease, it can be stated with the help of this table the total average cost of system is highly sensitive with respect to rate of interest earned
- Table 8.8 shows that when unit holding cost increases then total cost also increases and when holding cost decreases then total cost also decreases, it means we can say that holding cost and total cost both are directly proportional.
- Table 8.9 gives that demand rate is inversely proportionate to total cost and directly proportionate to time. When demand of product increases with time then total cost of a product decreases due to mass production, similarly when demand decreases, total production cost per unit increases.
- Table 8.10 shows that variation in constant part of holding cost (g) doesn't affect the total cost.

TABLE 8.7
Sensitivity Analysis for Variations in Interest Earned Rate (I_e)

I_e	Total Cost (TC_1)	Time (T)
+20%	3176	1.11
+15%	3259	1.52
+ 10%	3289	2.05
+5%	3375	2.35
0%	3425	2.57
−5%	3451	2.75
−10%	−3691	2.99
−15%	3746	3.05
−20%	−3865	3.22

TABLE 8.8
Sensitivity Analysis for Variations in Holding Cost (A)

A	Total Cost (TC_1)	Time (T)	Time (t_1)
+20%	755.11	3.213	1.625
+15%	735.561	3.141	1.595
+10%	705.451	3.052	1.583
+5%	688.068	2.931	1.531
0%	699.931	2.803	1.475
−5%	650.901	2.668	1.415
−10%	630.924	2.525	1.350
−15%	−625.899	2.416	1.275
−20%	610.500	2.344	1.118

TABLE 8.9
Sensitivity Analysis for Variations in Coefficient of Price in Demand Rate (b)

b	Total Cost (TC_1)	Time (T)	Time (t_1)
+20%	601.002	3.005	1.577
+15%	611.012	2.983	1.547
+10%	628.076	2.923	1.527
+5%	649.088	2.861	1.500
0%	699.931	2.803	1.475
−5%	690.614	2.749	1.451
−10%	711.148	2.697	1.428
−15%	737.211	2.625	1.415
−20%	757.156	2.605	1.399

TABLE 8.10
Sensitivity Analysis for Variations in Constant Part of Holding Cost (g)

g	Total Cost (TC_1)	Time (T)	Time (t_1)
+20%	679.385	2.511	1.205
+15%	677.501	2.651	1.351
+10%	675.575	2.770	1.420
+5%	672.804	2.786	1.447
0%	699.931	2.803	1.475
−5%	666.951	2.821	1.504
−10%	663.858	2.839	1.535
−15%	−662.412	2.849	1.555
−20%	−651.15	2.877	1.593

TABLE 8.11
Sensitivity Analysis for Time Varying Holding Cost (h)

h	Total Cost (TC_1)	Time (T)	Time (t_1)
+20%	671.103	2.785	1.442
+15%	670.911	2.790	1.458
+10%	670.611	2.795	1.464
+5%	670.273	2.799	1.470
0%	699.931	2.803	1.475
−5%	669.273	2.807	1.480
−10%	675.575	2.811	1.486
−15%	−684.654	2.816	1.492
−20%	697.116	2.821	1.499

- Table 8.7 shows that variation in varying holding cost (h) doesn't affect any significant difference in total cost.
- Table 8.7 shows that variation in inflation rate (r) directly proportionate to total cost. When inflation increases cost also increases and when inflation decreases total cost also decreases.
- Table 8.7 shows that variation in interest paid rate (I_p) directly proportionate to total cost and it is obvious.
- Table 8.7 shows that variation in interest earned rate (I_e) inversely proportionate to total cost and it is obvious. When interest earn increases it added in profit and that's why total cost decreases per unit. Similarly when interest earn decreases it subtracts from profit and that's why total cost increases per unit.

TABLE 8.12
Sensitivity Analysis for Variations in Inflation Rate (r)

r	Total Cost (TC_1)	Time (T)	Time (t_1)
+20%	785.11.	2.155	1.033
+15%	759.905	2.432	1.133
+10%	709.915	2.687	1.234
+5%	699.925	2.765	1.345
0%	657.931	2.803	1.475
−5%	599.944	2.955	1.543
−10%	565.123	2.968	1.654
−15%	−499.897	2.973	1.753
−20%	−448.456	2.999	1.854

TABLE 8.13
Sensitivity Analysis for Variations in Interest Paid Rate (I_p)

I_p	Total Cost (TC_1)	Time (T)	Time (t_1)
+20%	751.735	2.725	1.432
+15%	738.674	2.754	1.445
+ 10%	723.892	2.786	1.452
+5%	712.675	2.799	1.467
0%	699.931	2.803	1.475
−5%	658.345	2.812	1.482
−10%	−643.250	2.827	1.489
−15%	−637.875	2.836	1.492
−20%	697.116	2.821	1.499

TABLE 8.14
Sensitivity Analysis for Variations in Interest Earned Rate (I_e)

I_e	Total Cost (TC_1)	Time (T)	Time (t_1)
+20%	−622.870	2.136	1.005
+15%	−635.640	2.342	1.032
+ 10%	657.978	2.456	1.279
+5%	686.720	2.768	1.345
0%	699.931	2.803	1.475
−5%	709.342	2.945	1.670
−10%	717.890	3.034	1.749
−15%	724.345	3.259	1.790
−20%	742.670	3.645	1.845

8.10 CONCLUSION

In this model we have developed two inventory models: one is without shortages and another model with partial backlogged shortages with the realistic demand and varying carrying cost. Variable unit purchase cost underneath an order sizing base reduction environment in a seller viewpoint under the inflationary condition when the permissible delay in payment is allowed to the retailer within the given credit period. It is significant to mention that the inventory model is more economical with reduction in terms of cost optimization.

REFERENCES

Abad, P. L. "Optimal pricing lot sizing under condition of permissibility partial backordering". *Management Science* 42(8), 1093–1105 (1996).

Chang, C.-T., Chen, Y.-J., Tsai, Tzong-Ru, and Wu, Shuo-Jye. "Inventory models with stock- and price dependent demand for deteriorating items based on limited shelf space". *Yugoslav Journal of Operations Research*, 20, 55–69 (2010). doi:10.2298/YJOR1001055D.

Chang, C.T., Goyal, S.K., and Teng, J.T. "On an EOQ model for perishable items under stock-dependent selling rate and time-dependent partial backlogging". *European Journal of Operations Research*, 174(2), 923–929 (2006a).

Chang, H.J. and Lin, W.F. "A partial backlogging inventory model for non-instantaneous deteriorating items with stock-dependent consumption rate under inflation". *Yugoslav Journal of Operations Research*, 1, 35–54 (2010).

Chang, H.J., Teng, J.T., Ouyang, L.Y., and Dye, C.Y. "Retailer's optimal pricing and lot-sizing policies for deteriorating items with partial backlogging". *European Journal of Operations Research*, 168(1), 51–64 (2006b).

Chung, K. J., and Huang, T. S. "The optimal retailer's ordering policies for deteriorating items with limited storage capacity under trade credit financing." *International Journal of Production Economics*, 106, 127–145 (2007).

Dash, B., Singh, T., and Pattnayak, H. "An inventory model for deteriorating items with exponential declining demand and time-varying holding cost". *American Journal of Operations Research*, 04, 1–7, (2004). doi:10.4236/ajor.2014.41001.

Elsayed, E.A. and Teresi, Christina. "Analysis of inventory systems with deteriorating items". *International Journal of Production Research*, 21(4), 449–460 (2007).

Goel, A. and Pandey, R.K. "Production policy for inventory system with ramp type demand with shortages under inflation". *International Transactions in Applied Science*, 3(4), 1015–1029 (2011).

Goel A. and Pandey, R.K. "Flexible manufacturing inventory model with permissible delay in payment and cash discount under the effect of time value of money and inflation". *International Transactions in Mathematical Science and Computers*, 5(1), 131–144 (2012).

Harit, A., Sharma, A., and Singh, S.. "Effect of preservation technology on optimization of two warehouse inventory model for deteriorating". *Decision Analytics Applications in Industry*, Ch. 35, 431–442 (2020).

Khanra, S., Ghosh, S.K., and Chaudhuri, K.S. "An EOQ model for a deteriorating item with time dependent quadratic demand under permissible delay in payment". *Applied Mathematics and Computation*, 218, 1–9 (2011).

Kumar, S., Singh, Y., and Malik, A.K. "An inventory model for both variable holding and sales revenue cost". *Asian Journal of Management*, 8(4), 1111–1114 (2017).

Lee, C.C. and Hsu, S.L. "A two warehouse production model for deteriorating inventory items with time dependent demands," *European Journal of Operational Research*, 194(3), 700–710 (2009).

Liu, X., Yang, T., Pei, J., Liao, H., and Pohl, E.A. "Replacement and inventory control for a multi-customer product service system with decreasing replacement costs". *European Journal of Operational Research, Elsevier*, 273(2), 561–574 (2019).

Montgomery, D.C., Bazaraa, M.S., and Keswani, A.K. "Inventory models with a mixture of backorders and lost sales". *Naval Research Logistics*, 20/2, 255–263 (1973).

Ouyang, L.Y. Wu, K.S. and Cheng, M.C. "An inventory model for deteriorating items with exponential declining demand and partial backlogging". *Yugoslav Journal of Operations Research*, 15(2), 277–288 (2005).

Papachristos, S. and Skouri, K. "Optimal replenishment policy for deteriorating items with time-varying demand and partial-exponential type – backlogging". *Operations Research Letter*, 27, 175–184, (2000). doi:10.1016/S0167-6377(00)00044-4.

Ruidas, S., Seikh, M.R., Nayak, P.K., and Sarkar, B. (2019). "A single period production inventory model in interval environment with price revision". *International Journal of Applied and Computational Mathematics*, 5(1), 7 (2019).

Sekar, T. and Uthaya Kumar, R. "A production inventory model for single vendor single buyer integrated demand with multiple production setups and rework". *Uncertain Supply Chain Management*, 6(1), 75–90 (2018).

Shahabi, M., Tafreshian, A., Unnikrishnan, A., and Boyles, S.D. "Joint production–inventory–location problem with multi-variate normal demand". *Transportation Research Part B: Methodological, Elsevier*, 110(C), 60–78 (2018).

Singh, S.R., Gupta, V, and Goel, A. "An EOQ model with preservative technology investment when demand depends on selling price and credit period under two level of trade credit". *Procedia Technology*, 10, 227–235 (2013).

Singh, S.R. and Malik, A. K. "Two warehouses model with inflation induced demand under the credit period" *International Journal of Applied Mathematical Analysis and Applications*, 4(1), 59–70 (2009).

Skouri, K. and Papachristos, S. "Four inventory models for deteriorating items with time varying demand and partial backlogging A cost comparison." *Optimal Control Applications and Methods*, 24(6), 315–330 (2003a).

Skouri, K. and Papachristos, S. "Optimal stopping and restarting production times for an EOQ model with deteriorating items and time dependent partial backlogging". *International Journal of Production Economics*, 81–82, 525–531 (2003b).

Soni, H. and Shah, N.H. "Optimal ordering policy for stock-dependent demand under progressive payment scheme". *European Journal of Operational Research*, 184, 91–100 (2008).

Vashistha, V., Tomar, A., Soni, R., and Malik, A.K. "An inventory model for maximum lifetime products under the price and stock dependent demand rate". *International Journal of Computer Applications*, 132(15), 32–36 (2015).

Wee, H. M. "Economic production lot size model for deteriorating item with partial back ordering". *Computer and Industrial Engineering*, 24, 449–458 (1993).

Wee, H. M. "Deteriorating inventory model with quantity Discount, pricing and partial backordering". *I.J.P.E.*, 59, 511–518 (1999).

Zangwill, W.I. "A deterministic multi-period production scheduling model with backlogging". *Management Science*, 13, 105–119 (1966).

9 A Fuzzy Inventory Model for Non-Instantaneous Oxidizing Items with a Nonlinear-Hexagonal Fuzzy Number under the Effect of Learning

Neelanjana Rajput, R. K. Pandey, Anand Chauhan, and Bhuwan Chandra Joshi

CONTENTS

9.1 INTRODUCTION

Inventory optimization helps to deal with the effects of cost and about profit. With optimization technique, manufacturers earn minimum cost and maximum profit for their productions. The learning effect is the process (Figure 9.1) of the study, which cultivates the increases and decreases in production. Learning effect is a positive or negative effect of any production in the market, which emerge after some cycle time. Generally, the cost parameters are the important for any inventory model or industry, but in real life, there is some uncertainty in the nature of these parameters.

DOI: 10.1201/9781003089636-9

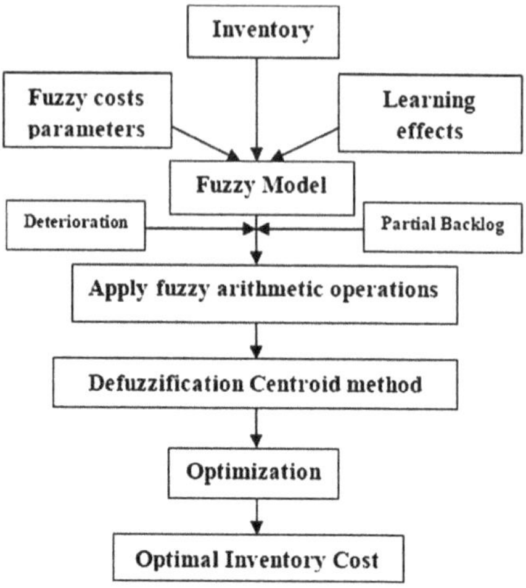

FIGURE 9.1 Flow chart of inventory with effect of learning.

So it is meaningful to take these as a fuzzy number. By the use of fuzzy set theory, the approximate result can be found and this will thereby optimize the inventory mode. The centroid defuzzification method is used here to convert the fuzzy model and optimize it. The optimal (minimum) inventory cost is found for the presented model.

9.2 LITERATURE REVIEW

Firstly, the father of fuzzy set theory, Zadeh [1] introduced the concept of the fuzzy system and developed some application of fuzzy set theory, which is used in inventory models. His contribution offers some useful insights for further research and it is very easy to deal with an uncertain nature of parameters through the use of fuzzy set theory. After that, D. Dubois et al. [2] introduced their research work on the algebraic operations on fuzzy numbers and the applications of these operations. There operations on real numbers convert into fuzzy numbers with the help of a fuzzification technique. In the field of inventory models, J. Kacprzyk et al. [3] developed an article about the long-term policy of fuzzy inventory model with replenishment norms. They deal with fuzzy demand and fuzzy replenishment in inventory. Then they able to make the decision for fuzzy inventory total cost function and optimize it. The concept of learning effect for a production introduced by J.C. Fisk et al. [4]. They discussed the conditions for bounded and unbounded learning for an optimal production lot size. They deal with the positive and negative learning effects of different parameters in production. M.Y. Jaber et al. [5] developed an inventory model with learning effects on the production lot size for both the finite and infinite tie cycle. They deal with economic manufactured quantity (EMQ) model with the time of interruption period and then minimize the total inventory cost with the help of a numerical example. The concept of the stock-dependent demand is discussed by C.Y. Dey [6]. In his research work, he studied the inventory model for deteriorating items with partial backlogging with permissible delay in payment. After all of these factors, he finally obtained the optimal result for his model and optimizes the inventory cost. Sometime later, M.Y. Jaber [7] discussed the effect of learning on the forgetting inventory models and the further applications of the research work. Similarly, M.Y. Jaber et al. [8] developed an EPQ (economic production quantity) model with the concept of learning effects. They learnt the effects of learning on the production model with all the parameters and got the optimal inventory cost for the production model.

J.C.-H. Pan et al. [9] investigated an inventory model with the effects of learning on the setup cost. They studied the effects of the mixture of the partial lost sales and backorder in their inventory model. The main result of their model is to minimize the total annual cost by optimization techniques and a suitable solution algorithm. The new concept of remanufacturing model is introduced by M.Y. Jaber et al. [10]. They worked on an economic production model with learning effects and remanufacturing the products and obtained higher profits for the industry. The remanufacture of the items helps the manufacturer for the shortage in inventory. N. Kumar et al. [11] studied an inventory model with optimal replenishment policy for the deteriorating products under two-level shortages. They also discussed the effects of learning with partial backlog in the inventory. They minimize the total inventory cost through optimization methods. For the deteriorating items, Y. Tan et al. [12] introduced an optimal stochastic inventory control model under partial backlogging. They discuss an inventory model for a finite time horizon. In their study, they consider the unsatisfied demand and fulfilled it immediately before the next demand arrives and then they studied the effect of learning due to partial backlogging. S.R. Singh et al. [13] defined the production inventory model for preservation and investment under the learning effect. They also deal with the shortage of inventory and exponential rate of backlogging. They found the optimal inventory cost in two cases; for the model which ends with shortages and for the model which starts with the shortages. They show the learning effect due to the set-up and production costs for the model. A.K. Goyal et al. [14] developed an economic order quantity model where the demand rate is dependent on the selling price and for some deterioration rate. They also discussed the learning effects on different costs. Specially, they optimize the vendor's total cost for the model. The learning effect is now useful in recent study about the industry 4.0-based manufacturing system, Yaping Fu et al. [15] developed an inventory model about the two-objective stochastic flow shop system under industry 4.0-based manufacture. They deal with the deteriorating items and use the advanced machines, which have communications, self-training behaviors and self-optimization techniques. Rajput et al. [16] proposed an inventory model with a different type of demand function, discussed the importance of fuzzy parameters in healthcare industries. They used triangular fuzzy number for the demand and defuzzified the model with signed distance method and get the maximum profit for all three models. Under the same topic of the industry 4.0-based manufacturing system, Yaping Fu et al. [17]

introduced an artificial-molecule chemical optimization model for deteriorating chemical items and then they studied the learning effects of the items in the scheduling problem.

Rajput et al. [18] developed the inventory model with fuzzy demand rate. They used signed distance method for defuzzification and get the optimal inventory cost. They used triangular fuzzy number as well as defuzzification by signed distance method. In the field of fuzzy inventory models, D. Khatua et al. [19] developed a fuzzy optimal inventory model of the products; they also discussed the fuzzy effect of learning on the model with finite cycle time. They consider all the parameters are fuzzy numbers and solved the total profit function by the Runge-Kutta forward-backward method in MATLAB® software. In the fuzzy environment, Rajput et al. [20] proposed an inventory model for both the crisp and the fuzzy environments. They included reliability induced fuzzy demand Graded Mean Integration method for defuzzification for the cost function. With the fuzzy set theory, there are several applications in inventory models like S. Maity et al. [21] developed their research article on the comprehensive study of an EOQ model with nonlinear heptagonal dense fuzzy number under partial backlogging. They apply both the symmetric and asymmetric heptagonal fuzzy number and developed a new ranking method for defuzzification for this fuzzy model. In the recent research work, J.B. Wang et al. [22] introduced their article for the job release scheduling under learning effect. The main motive of their research is to minimize the total completion time with lower bound on the objective function. They use the heuristic algorithm for the evaluation of the optimal result. The present study introduced an inventory model with the learning effect policy. In this situation, the uncertainty of the demand rate has been solved with the application of nonlinear asymmetric hexagonal-fuzzy number and the centroid defuzzification method for this fuzzy model. The introduction and literature review are in Sections 9.1 and 9.2, some basic definitions related to fuzzy number and centroid method are in Section 9.3. In Section 9.4, we introduce some assumptions and notations for the model. Mathematical modeling is described in Section 9.5, Section 9.6 represents the algorithm for optimality criteria, and numerical illustration example shows in section 15.7 with a 3-d graph of objective function and the sensitivity analysis is shown in section 15.8 with a suitable table and bar chart. We construct the conclusion for the application of fuzzy theory in inventory models in Section 9.9.

9.3 PRELIMINARY CONCEPTS

Definition 1- Fuzzy Set: - A fuzzy set $\tilde{A}$ can be defined as

$$\tilde{A} = \left\{ \left(a, \mu_{\tilde{A}}(a) \right) : a \in A \right\}$$

Where $\mu_{\tilde{A}}(a)$ is the degree of membership function of a in A and range of its value is 0 to 1, i.e. $\mu_{\tilde{A}}(a) \in [0,1]$.

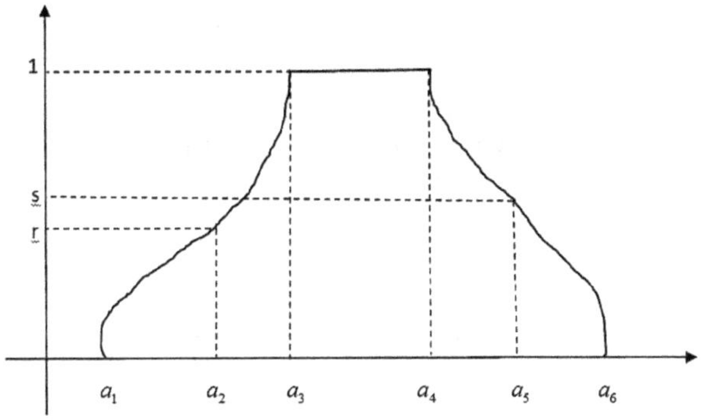

FIGURE 9.2 Graphical presentation of Nonlinear Hexagonal fuzzy number.

Definition 2 - Fuzzy Number: - Let z be the elements of a crisp set Z. The membership function of in a crisp subset S of Z is written as $\mu_{\tilde{S}}(z)$

$$\mu_{\tilde{S}}(z) = \begin{cases} 1, & z \in \tilde{S} \\ 0, & \text{else} \end{cases}$$

Definition 3 - Nonlinear Hexagonal Fuzzy Number with Asymmetry: - Consider the fuzzy set defined on R, which is called a nonlinear hexagonal fuzzy number (in Figure 9.2) if the membership function of $\tilde{A} = (a_1, a_2, a_3, a_4, a_5, a_6: r, s)$ is given as

$$\mu_{\tilde{A}(x)} = \begin{cases} r\left(\dfrac{x-a_1}{a_2-a_1}\right)^{n_1} & \text{, if } a_1 \le x \le a_2 \\ 1-(1-r)\left(\dfrac{x-a_2}{a_3-a_2}\right)^{n_2} & \text{, if } a_2 \le x \le a_3 \\ 1 & \text{, if } a_3 \le x \le a_4 \\ 1-(1-s)\left(\dfrac{a_5-x}{a_5-a_4}\right)^{m_1} & \text{, if } a_4 \le x \le a_5 \\ s\left(\dfrac{a_6-x}{a_6-a_5}\right)^{m_2} & \text{, if } a_5 \le x \le a_6 \\ 0 & \text{, if } x > a_6 \end{cases}$$

Definition 4 - Centroid Method:- Let $\bar{A}$ be a q-fuzzy number. The defuzzification of $\bar{A}$ with centroid method is given by

$$\frac{\int x\mu_{\bar{A}(x)dx}}{\int \mu_{\bar{A}(x)dx}}$$

$\mu_{\bar{A}(x)}$ is taken from the definition (3). Let A^* be the centroid value of nonlinear hexagonal fuzzy no $\tilde{A}$, then

$$A^* = \frac{\int x\mu_{\bar{A}(x)dx}}{\int \mu_{\bar{A}(x)dx}}$$

$$= \frac{\int_{a_1}^{a_2} x\mu(x)dx + \int_{a_2}^{a_3} x\mu(x)dx + \int_{a_3}^{a_4} x\mu(x)dx + \int_{a_4}^{a_5} x\mu(x)dx + \int_{a_5}^{a_6} x\mu(x)dx}{\int_{a_1}^{a_2} \mu(x)dx + \int_{a_2}^{a_3} \mu(x)dx + \int_{a_3}^{a_3} \mu(x)dx + \int_{a_4}^{a_5} \mu(x)dx + \int_{a_5}^{a_6} \mu(x)dx} \quad (9.1)$$

Definition 5 - Arithmetic Operations on fuzzy numbers:-

Suppose $\bar{\tau} = (\tau_1, \tau_2, T_3, T_4, T_5, T_6)$ and $\bar{\varsigma} = (\varsigma_1, \varsigma_2, \varsigma_3, \varsigma_4, \varsigma_5, \varsigma_6)$ are two triangular fuzzy numbers, then arithmetical operations are defined as;

$$\text{Addition: } \bar{\tau} \oplus \bar{\zeta} = (\tau_1 + \varsigma_1, \tau_2 + \varsigma_2, \tau_3 + \varsigma_3, \tau_4 + \varsigma_4, \pi_5 + \varsigma_5, \pi_6 + \varsigma_6)$$

$$\text{Subtraction: } \infty\bar{\tau} \odot \bar{\zeta} = (\tau_1 - \varsigma_6, \tau_2 - \varsigma_5, T_3 - \varsigma_4, \tau_4 - \varsigma_3, \tau_5 - \varsigma_2, \tau_6 - \varsigma_1)$$

$$\text{Multiplication: } \bar{\tau} \otimes \bar{\zeta} = (\tau_1\varsigma_1, \tau_2\varsigma_2, \tau_3\varsigma_3, \tau_4\varsigma_4, \tau_5\zeta_5, \tau_6\zeta_6)$$

1. For any real number $K, K\bar{\tau} = (K\tau_1, K\tau_2, K\tau_3, K\tau_4, K\tau_5, K\tau_6)$ if $K > 0$ and $K\bar{\tau} = (K\tau_6, K\tau_5, K\tau_4, K\tau_3, K\tau_2, K\tau_1)$ if $K < 0$.
2. If $\tau_1, \tau_2, T_3, \tau_4, \tau_5, \tau_6, \varsigma_1, \varsigma_2, \varsigma_3, \varsigma_4, \varsigma_5, \varsigma_6$ are all nonzero positive real numbers, then division of two fuzzy number is

$$\bar{\tau} \oslash \bar{\xi} = \left(\frac{\tau_1}{\varepsilon_6}, \frac{T_2}{\varsigma_5}, \frac{\tau_3}{\varsigma_4}, \frac{\tau_4}{\xi_3}, \frac{\tau_5}{\varsigma_2}, \frac{\tau_6}{\S_1} \right).$$

9.4 ASSUMPTION AND NOTATION

9.4.1 Assumptions for the Model

- There is a stock-dependent demand rate, which is follows as

$$d(t) = \alpha + \beta I(t), I(t) > 0 \,\&\, \alpha > 0, 0 < \beta \langle 1 \text{ with } \alpha \rangle \beta$$

- There are partially backlogged shortages of inventory, which is follows as

$$N(t) = e^{-\delta(T-t)}, 0 < \delta < 1$$

- The holding cost is initially constant as h and partly decreasing due to learning effect of workers, which is given as $\left[\frac{h_0}{m^6} + h\right], b > 0$.
- The deterioration rate is dependent on time.

- The length of the cycle is T.
- The lead time is to be considered as zero for this model.

Notations for the model

α	demand parameters.
η	rate of deterioration.
t_0	time, when the product has no deterioration.
i_m	inventory at its maximum level.
h	initial holding cost per cycle.
A	fuzzy deterioration cost per unit.
$\tilde{B}$	fuzzy shortage cost per unit.
$\tilde{CC}$	fuzzy lost sale cost per unit.
$\tilde{O}$	fuzzy ordering cost per unit.
$\tilde{D}$	fuzzy purchasing cost per unit.
$\widetilde{PRIC}$	Total relevant inventory cost in fuzzy model.
D	fuzzy purchasing cost per unit
$\widetilde{TRIC}$	Total relevant inventory cost in fuzzy model

9.5 MATHEMATICAL MODEL IN FUZZY ENVIRONMENT

Figure 9.3 shows the inventory level (I (t)) with respect to time (t). The cycle time T divides into three sub-intervals. In the interval [0, t0], inventory depends on the demand which depends on the stock. The inventory goes to zero in the next interval [t0, t1] due to demand and the rate of deterioration, and in the last time interval [t1, T]

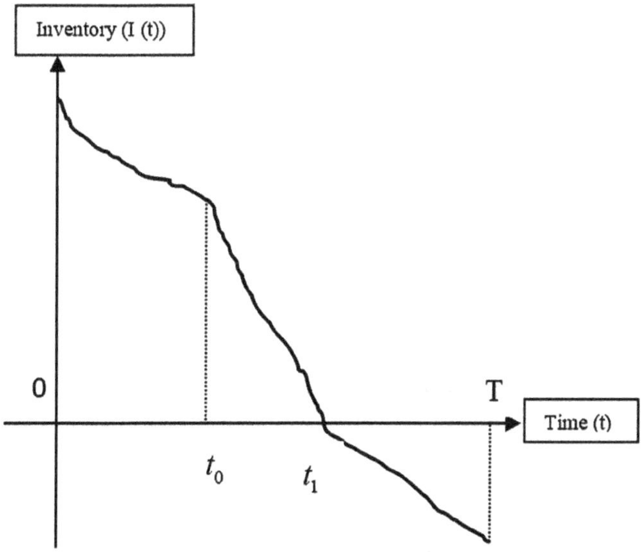

FIGURE 9.3 Behavior of inventory with time.

the shortage starts due to partial backlogging and then some lost sales occur. For each time interval, the inventory system is also depicted in the below equations:

$$\frac{dI_1}{dt} = -\left[\alpha + \beta I_1(t)\right]; 0 \leqslant t \leqslant t_0$$
$$\frac{dI_2}{dt} + \eta t I_2(t) = -\left[\alpha + \beta I_2(t)\right]; t_0 \leqslant t \leqslant t_1 \tag{9.2}$$
$$\frac{dI_3}{dt} = -\left[\alpha e^{-\delta(T-t)}\right]; t_1 \leqslant t \leqslant T$$

With the conditions, $I_1(0) = i_m$, $I_2(t_1) = 0$ & $I_3(t_1) = 0$ and we get the solutions of the above differential Equation (9.2) below

$$I_1 = e^{-\beta t}\left[-\alpha\left(\frac{\beta t^2}{2} + t\right) + i_m\right]$$
$$I_2 = \alpha\left[\frac{1}{6}\eta\left(t_1^3 - t^3\right) + \frac{1}{2}\beta\left(t_1^2 - t^2\right) + \left(t_1 - t\right)\right] e^{-\left(\beta t + \frac{b^2}{2}\right)} \tag{9.3}$$
$$I_3 = \alpha\left[t_1 - \delta\left(t_1 T - \frac{t_1^2}{2}\right)\right] - \alpha\left[t - \delta\left(tT - \frac{t^2}{2}\right)\right]$$

Let us take the continuity of inventory graph at time t_0 in Equation (9.3), we have $I_1(t_0) = I_2(t_0)$ and then,

$$e^{\beta t_0}\left[-\alpha\left(\frac{\beta t_0^2}{2} + t_0\right) + i_m\right] = \alpha\left[\frac{1}{6}\eta\left(t_1^3 - t_0^3\right) + \frac{1}{2}\beta\left(t_1^2 - t_0^2\right) + \left(t_1 - t_0\right)\right] e^{-\left(s t_0 + \frac{n_0^2}{2}\right)}$$

Now, for each cycle period, the maximum level q_1(or i_m) will become

$$q_1 = i_m = \alpha e^{\beta t_0}\left[m\left(t_1 - t_0\right)t_0 + \frac{1}{4}\left(t_0^2 - t_1^2\right)t_0^2(\beta\eta) + \frac{1}{6}\left(t_1^3 - t_0^3\right)t_0(\beta\eta) + \right.$$
$$\left.\frac{1}{2}\beta\left(t_1^2 - t_0^2\right) + \frac{1}{2}\eta\left(t_0 - t_1\right)t_0^2 + \frac{1}{6}\eta\left(t_1^3 - t_0^3\right) + \left(t_1 - t_0\right)\right] + \alpha t_0\left(1 - \frac{\beta t_0}{2}\right) \tag{9.4}$$

For the maximum level of backlogging items in each cycle, replace $t = T$ in $I_3(t)$ equation and get

$$q_2 = -I_3(t) = \alpha\left[t_1 - \delta\left(t_1 T - \frac{t_1^2}{2}\right)\right] - \alpha\left(T - \frac{\delta T^2}{2}\right) \tag{9.5}$$

Thus from Equations (9.4) and (9.5), the total quantity for the order will become

$$
\begin{aligned}
Q = q_1 + q_2 = \alpha e^{\beta t_0}\Big[& m(t_1 - t_0)t_0 + \frac{1}{4}(t_0^2 - t_1^2)t_0^2(\beta\eta) + \frac{1}{6}(t_1^3 - t_0^3)t_0(\beta\eta) \\
& + \frac{1}{2}\beta(t_1^2 - t_0^2) + \frac{1}{2}\eta(t_0 - t_1)t_0^2 + \frac{1}{6}\eta(t_1^3 - t_0^3) + (t_1 - t_0)\Big] \\
& + \alpha t_0\left(1 - \frac{\beta t_0}{2}\right) + \alpha\left[t_1 - \delta\left(t_1 T - \frac{t_1^2}{2}\right)\right] - \alpha\left(T - \frac{\delta T^2}{2}\right)
\end{aligned}
\tag{9.6}
$$

The fuzzy purchasing cost for the cycle period $[0, T]$ will be

$$
\begin{aligned}
\widetilde{PC} &= \tilde{D}(q_1 + q_2) \\
&= \tilde{D}\Big[\alpha e^{\beta t_0}\Big[m(t_1 - t_0)t_0 + \frac{1}{4}(t_0^2 - t_1^2)t_0^2(\beta\eta) + \frac{1}{6}(t_1^3 - t_0^3)t_0(\beta\eta) \\
&\quad + \frac{1}{2}\beta(t_1^2 - t_0^2) + \frac{1}{2}\eta(t_0 - t_1)t_0^2 + \frac{1}{6}\eta(t_1^3 - t_0^3) + (t_1 - t_0)\Big] \\
&\quad + \alpha t_0\left(1 - \frac{\beta t_0}{2}\right) + \alpha\left[t_1 - \delta\left(t_1 T - \frac{t_1^2}{2}\right)\right] - \alpha\left(T - \frac{\delta T^2}{2}\right)\Big]
\end{aligned}
\tag{9.7}
$$

The fuzzy ordering cost is given by

$$
\widetilde{OC} = \tilde{O} \tag{9.8}
$$

Due to the rate of deterioration of the product, the fuzzy deterioration cost will become

$$
\begin{aligned}
\widetilde{DC} &= \tilde{A}\left[I_2(t_0) - \int_{t_0}^{t_1}(\alpha + \beta I_2(t))dt\right] \\
&= \tilde{A}\Bigg[\alpha\left(1 - \frac{\beta}{\beta + \eta t_0}\right)\left(\frac{1}{2}\beta(t_1^2 - t_0^2) + \frac{1}{6}\eta(t_1^3 - t_0^3) + (t_1 - t_0)\right)e^{-\left(\beta t_0 + \frac{\eta t_0^2}{2}\right)} \\
&\quad + \frac{\alpha\beta\left(\beta t_0 + \frac{\eta t_0^2}{2} + 2\right)e^{-\left(\beta t_0 + \frac{\eta t_0^2}{2}\right)}}{(\beta + \eta t_0)(\beta + \eta t_0)} - \frac{\alpha\beta\left(\beta t_1 + \frac{\eta t_1^2}{2} + 2\right)e^{-\left(\beta t_1 + \frac{\eta t_1^2}{2}\right)}}{(\beta + \eta t_1)(\beta + \eta t_1)} + \alpha(t_1 - t_0)\Bigg]
\end{aligned}
\tag{9.9}
$$

Due to the learning effects, the holding cost for per cycle will be

$$\mathrm{HC}=\left(\frac{h_0}{m^b}+h\right)\left[-\frac{\alpha\left(\frac{1}{\beta}+1\right)}{\beta}-\frac{\alpha\left(\frac{1}{2}\beta\left(t_1^2-t_0^2\right)+\frac{1}{6}\eta\left(t_1^3-t_0^3\right)+\left(t_1-t_0\right)\right)\left(\alpha e^{-\left(\beta t_0+\frac{\eta t_0^2}{2}\right)}\right)}{\beta+\eta t_0}\right.$$

$$-\frac{\alpha\left(\beta t_0+\frac{\eta t_0^2}{2}+2\right)e^{-\left(\beta t_0+\frac{\eta t_0^2}{2}\right)}}{\left(\beta+\eta t_0\right)\left(\beta+\eta t_0\right)}+\frac{\alpha\left(\beta t_1+\frac{\eta t_1^2}{2}+2\right)e^{-\left(\beta t_1+\frac{\eta t_1^2}{2}\right)}}{\left(\beta+\eta t_1\right)\left(\beta+\eta t_1\right)}$$

$$\left.+\frac{\left(\left(\frac{\beta t_0^2}{2}+t_0\right)+\frac{\beta t_0+1}{\beta}+1\right)\left(\alpha e^{-\beta t_0}\right)}{\beta}\right] \tag{9.10}$$

$$\left(\frac{h_0}{m^b}+h\right)\left\{-\alpha e^{\beta t_0}\left(m\left(t_1-t_0\right)t_0+\frac{1}{4}\left(t_0^2-t_1^2\right)t_0^2\left(\beta\eta\right)+\frac{1}{6}\left(t_1^3-t_0^3\right)t_0\left(\beta\eta\right)+\right.\right.$$
$$\left.\left.\frac{1}{2}\beta\left(t_1^2-t_0^2\right)+\frac{1}{2}\eta\left(t_0-t_1\right)t_0^2+\frac{1}{6}\eta\left(t_1^3-t_0^3\right)+\left(t_1-t_0\right)\right)-\alpha t_0\left(1-\frac{\beta t_0}{2}\right)\right\}$$

Due to partial backlog in the interval $[t_1, T]$ the fuzzy lost sales cost is

$$\widetilde{LSC}=\tilde{C}\int_{t_1}^{T}\alpha e^{-\delta(T-t)}dt=C\frac{1}{2}\left(T-t_1\right)^2\left(\alpha\delta\right) \tag{9.11}$$

The fuzzy cost for the shortage items is given by

$$\widetilde{SC}=\tilde{B}\left[\frac{\alpha}{2}\left(T^2-t_1^2\right)-\alpha\delta\left(-t_1T^2+t_1^3T+\frac{5t_1^3}{6}+\frac{T^3}{6}\right)+\alpha t_1T\right] \tag{9.12}$$

Now, from Equations (9.7) to (9.12), the total relevant inventory cost in fuzzy environment will be

$$\widetilde{PRIC}=\frac{1}{T}\left[\widetilde{PC}+\widetilde{OC}+\widetilde{DC}+HC+\widetilde{LSC}+\widetilde{SC}\right] \tag{9.13}$$

Due to fuzzy cost parameters in the above Equation (9.13), first we solve it by fuzzy operations (Definition 5) then the defuzzification by Centroid method (Definition 4), now we will use optimality criteria (described in next section) to get optimal (minimal) result.

9.6 ALGORITHM FOR OPTIMALITY CRITERIA

To find the minimum value of the $\widetilde{PRIC}$ with respect to t_1, we have to follow these steps:

Step 1 Start with $\frac{\widetilde{PRIC}}{dt_1} = 0$ and get t_1^*.

Step 2 Using t_1^* which found in step 1, evaluate the value of $\widetilde{PRIC}\left(t_1^*\right)$ from the cost function Equation (9.13).

Step 3 If the second derivative of $\widetilde{PRIC} > 0$ for the critical value t_1^*. Then, we get the minimum value of cost function which is a convex function.

Step 4 Repeat step 2 and step 3 until we get the optimal solution for the cost function.

9.7 NUMERICAL ILLUSTRATION

To satisfy the result of the above fuzzy model, we consider α = 150, β = 0.44, $\eta = 0.2$, $t_0 = 3.5$, $h = 5$, $h_0 = 2$, $\tilde{A} = (0.5, 1, 1.2, 2, 2.5, 3)$, $B = (0.8, 1.5, 2, 2.4, 3, 3.2)$, C=(0.4,1,1.6,2.2,2.5,3), $\tilde{O} = (170, 176, 190, 195, 199, 205)$, $\tilde{D} = (2, 2.5, 3, 3.8, 4.5, 5)$, T = 10 days, m = 3, b = 0.1. From the Equation (9.13), we get the optimal total cost will be $\widetilde{PRIC} = 35192.6$ $\widetilde{\text{PRIC}} = 35192.6$ at the time $t_1 = 2.1033$.

Figure 9.4 shows that the convex character of inventory cost function in fuzzy environment. It is clear that, the cost has a minimum point at $t_1 \approx 2$ days; the

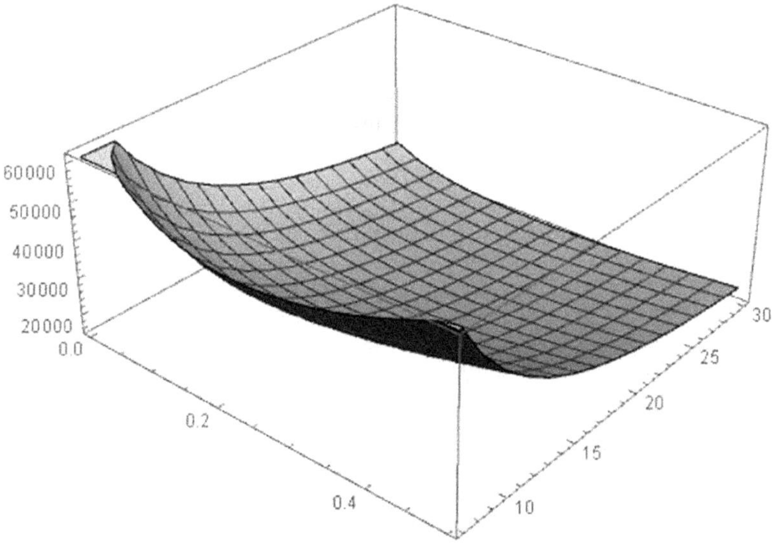

FIGURE 9.4 Convex presentation of the total cost function.

inventory cost is $35192.6 for fuzzy model with the use of fuzzy techniques. With the help of fuzzy techniques, the inventory cost is optimize and got best result for the example.

9.8 SENSITIVITY ANALYSIS

The consideration of a sensitivity analysis for the fuzzy total inventory cost with the parametric changes from –50% to +50% of the parameters α, β, η, δ and b respectively in Table 9.1.

From Table 9.1, it is seen that the parameters α is highly sensitive compared to all other parameters β, η, δ and b. while there is almost no effect on the inventory cost due to change in the δ. The change of α and β at –50% and +50% the average inventory cost increases. With this change in η and in b at –50% and +50% the objective cost function decreases slightly.

From Figure 9.5, it is clear that the change in inventory cost due to demand parameter α is highly sensitive compared with other parameters in the fuzzy model. While the δ doesn't have an effect on fuzzy inventory cost. The change in β, η and in b effects on the fuzzy inventory cost minor. In the case of β and η, we got no feasible solution for the cost function in some cases for negative changes in the parameter.

TABLE 9.1
Sensitivity Analysis for the Fuzzy Model

Parameters	change	t_1	$\widetilde{PRIC}$
α	75	2.10168	9541.49
	112.5	2.10279	20353.3
	187.5	2.1037	54062.5
	225	2.10393	76963
β	0.22	No feasible solution	
	0.33	No feasible solution	
	0.55	1.71599	39884.9
	0.66	1.46476	43685.1
η	0.1	No feasible solution	
	0.15	No feasible solution	
	0.25	1.87299	34323.6
	0.3	1.73115	33279.3
δ	0.005	2.10285	35189.4
	0.0075	2.1031	35191
	0.0125	2.10361	35194.2
	0.015	2.10387	35195.7
b	0.05	2.10386	35698.3
	0.075	2.10361	35442
	0.125	2.10311	34950
	0.15	2.10287	34713.9

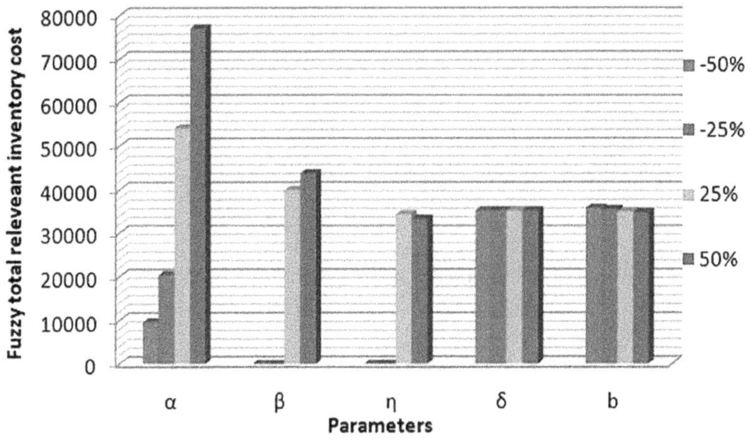

FIGURE 9.5 Graphical presentation of change $\widetilde{PRIC}$ in due to change in fuzzy parameters.

9.9 CONCLUSION

The chapter presents a fuzzy inventory model for non-instantaneous decaying products. The costs are fuzzy numbers and demand rate is stock-dependent, deterioration rate and partial backlogging occurs. The uncertain nature of cost shows the fluctuations of the market. To deal with this nature, we proposed here the defuzzification centroid method, and then get the optimal result as much as real. The effect of learning on holding cost shows, how the holding cost will decrease after some cycle time. An experimental numerical example the model as well as describe the analysis of the variation in the total inventory cost with different parameters.

REFERENCES

1. Zadeh, L.A.; Fuzzy sets. *Information and Control*, 8, (1965), 338–356.
2. Dubois, D. and Prade, H.; Operations on Fuzzy numbers. *International Journal of System Science*, 9(6), (1978), 613–626.
3. Kacprzyk, J. and Stanieski, P.; Long-term inventory policy-making through fuzzy decision-making models. *Fuzzy Sets and Systems*, 8(2), (1982), 117–132.
4. Fisk, J.C. and Ballou, D.P.; Production lot sizing under a learning effect. *AIIE Transactions*, 14(4), (1985), 257–264.
5. Jaber, M.Y. and Bonney, M.; The effect of learning and forgetting on the optimal lot size quantity (EMQ) of intermittent production runs. *Production Planning & Control: The Management of Operations*, 9(1), (1998), 20–27.
6. Dey, C.Y.; A deteriorating inventory model with stock-dependent demand rate and partial backlogging under conditions of permissible delay in payment. *OPSEARCH*, 39(3), (2002), 189–201.
7. Jaber, M.Y.; Learning and forgetting models and their applications. In: Badiru, A.B. (Ed.), *Handbook of Industrial and Systems Engineering*. CRC Press, Boca Raton, FL, 30, (2006), 1–27.
8. Jaber, M.Y., Goyal, S.K. and Imran, M.; Economic production quantity subject to learning effects. *International Journal of Production Economics*, 115(1), (2008), 143–150.

9. Pan, J. C.-H. and Lo, M.-C.; The learning effect on setup cost reduction for mixture inventory models with variable lead time. *Asia-Pacific Journal of Operational Research*, 25(4), (2008), 513–529.
10. Jaber, M.Y. and El Saadany, A.M.A.; An economic production and remanufacturing model with learning effects. *International Journal of Production Economics*, 131(1), (2011), 115–127.
11. Kumar, N., Singh, S.R. and Kumari, R.; Learning effect on an inventory model with two-level storage and partial backlogging under inflation. *International Journal of Services and Operations Management*, 16(1), (2013), 105–122.
12. Tan, Y. and Weng, M.X.; Optimal Stochastic inventory control with deterioration and partial backlogging/service-level constraints. *International Journal of Operational Research*, 16(2), (2013), 241–261.
13. Singh, S.R. and Rathore, H.; *Production-inventory model with preservation technology investment under the effect of learning and shortages. Proceedings of 3rd International Conference on Reliability, Infocom Technologies and Optimization*, (2014), 1–6.
14. Goyal, A.K. and Chauhan, A.; An EOQ model for deteriorating items with selling price dependent demand rate with learning effect. *Nonlinear Studies*, 23(4), (2016), 541–550.
15. Yaping, Fu., Ding, J., Wang, H. and Wang, J.; Two-objective stochastic flow-shop scheduling with deteriorating and learning effect in industry 4.0-based manufacturing system. *Applied Soft Computing*, 68, (2018), 847–855.
16. Rajput, N., Pandey, R.K., Singh, A.P. and Chauhan, A.; An optimization of fuzzy EOQ model in healthcare industries with three different demand patterns using signed distance technique. *Mathematics in Engineering, Science and Aerospace*, 10(2), (2019), 205–218.
17. Yaping, Fu., Mengchu, Z., Xiwang, G. and Liang, Qi.; Artificial-molecule-based chemical reaction optimization for flow shop scheduling problem with deteriorating and learning effects. *IEEE Access*, 7, (2019), 53429–53440.
18. Rajput, N., Singh, A.P. and Pandey, R.K.; Optimize the cost of a fuzzy inventory model with shortage using sign distance method. *International Journal of Research in Advent Technology*, 7(5), (2019), 198–202.
19. Khatua, D., Maity, K. and Kar, S.; A Fuzzy optimal control inventory model of product-process innovation and fuzzy learning effect in finite time horizon. *International Journal of Fuzzy Systems*, 21(5), (2019), 1560–1570.
20. Rajput, N., Chauhan, A. and Pandey, R.K.; Fuzzy EOQ model with reliability induced demand and defuzzification by Graded Mean Integration. In: Ram, Mangey (Ed.), *Recent Advances in Mathematics for Engineering*, Published March 31, 2020 by CRC Press (2020), 305–326.
21. Maity, S., Chakraborty, A., De, K., Mondal, S.P. and Alam, S.; A comprehensive study of a backlogging EOQ model with nonlinear heptagonal dense fuzzy environment. *RAIRO: Research Operationnelle*, 54, (2020), 267–286.
22. Wang, J.B., Gao, M., Wang, J.J., Lu, L. and He, H.; Scheduling with a position-weighted learning effect and job release dates. *Engineering Optimization*, 52(9), (2020), 1475–1493.

10 Optimal Analysis of Machine Interference Problem with Standby, Random Switching Failure, Vacation Interruption and Synchronized Reneging

Amit Kumar, Mohamed Boualem, Amina Angelika Bouchentouf, and Savita

CONTENTS

10.1 INTRODUCTION

This era has been labelled Industry 4.0, referring to a new phase in the industrial revolution. Industry 4.0 focuses strongly on machine learning, automation, real-time data, and internet of things (IoT) or smart technology, etc. Machine repair problems have a vital role in our real-life system: communication, manufacturing, printed

DOI: 10.1201/9781003089636-10

circuit board (PCB), and so on. In an active-redundancy, fault tolerance refers to a system capability (computer, cloud cluster, etc.) to continue operating without interruption when one or more of its components fails.

Random failure and systematic repairs of machining systems units significantly affect the productivity of the machining systems. Thus, appropriate maintenance and repair policies are necessary for the continuous and smooth operation of the systems. It is worth pointing out that no machining system operates at full efficiency and capacity if and/or when the machine in operation is liable to fail. This could have a very bad impact on industries' profits. This problem can be resolved by utilizing an appropriate set of standby machines and parts. For the proper working of the machining system, when a machine fails, for a range of reasons, it is immediately replaced with a standby component that may be available in the machining system itself. Once the machine is repaired, it either joins a group of standby machines, or goes back to the operation owing to the lack of working machines. Units of standby can be broadly described in three categories, namely, cold standby, hot standby, and warm standby.

10.2 LITERATURE REVIEW

In the automated manufacturing systems, operating machines are subject to random breakdowns which may result in a significant loss of production. To curtail the interludes instigated by operating machine breakdowns, the system may be supported by optimal preventive and maintenance policy. Under the preventive policy, spares provisioning is one of the good options at some additional cost in which available spare machine instantly replaces the failed operating machine whenever an operating machine breakdowns. There is always a chance of failure of switching process from spare state to operating state with significant probability.

Switching failure and fault tolerance machining systems have been the subject of widespread study (c.f. [1–4]). Shekhar et al. [5] investigated machine repair problems with fault tolerance, failure, and delays using the supplementary technique. Jain and Meena [6] studied the M/G/1 machining systems with imperfect fault coverage, recovery along with and replacement facility using the recursive algorithm. Hsu et al. [7] dealt with the multi-server machine repair problem with switching failure, repair pressure coefficient, and reboot delay.

A working vacation is a special kind of vacation policy in which the server provides the service at a slower rate. The idea was introduced by Servi and Finn [8]. Since then, many research works have been conducted on the subject. Laxmi and Rajesh [9] studied the single server queue model with variant working vacation and found the transient probability using the generating method. Meena et al. [10] investigated a machine repair problem with standby and vacation policy. Using recursive and supplementary variable techniques, steady-state probabilities were found; subsequently, some essential performance measures were derived. Patterson and Korzeniowski [11] proposed the M/M/1 model with a working vacation and unreliable service queueing. They found the stationary probabilities of the system, and obtained the closed form of the expected value as well as of the variance, using the Laplace transformation. Recently, Kumar [12] dealt with a single server queueing system with customer behavior and multiple vacations using the confluent hypergeometric function. Many researchers (c.f. [13–17]) worked on working vacation (WV) policy, while only a few focused on machine repair problems with spare and working vacation policy.

Under the optimal repair facility, during the working vacation, the repairman is a good choice for the system designer in which the repairman renders repair with lower rates rather than completely terminating the repair during a vacation period or avoiding the idle time of the repairman. The repairman begins working when there is no failed machine in the system. If the system is empty at the completion of the vacation, the repairman gets another working vacation; otherwise they switch to actual repair rates. The vacation period is interrupted once all available spare machines are exhausted and the system is functioning at a lower rate. This model is motivated by remote systems where failed machines have to wait for a certain repair facility to abandon the system whenever the repair facility visits the system. Then some of the failed machines decide whether or not to leave the system "synchronously". Synchronized departure in queueing systems were first introduced by Ivo et al. [18], where a single server queueing model with vacation and synchronized reneging has been examined, the transient probability of the system was found using the probability generation function (PGF). Then, Dimou and Economou [19] studied a single server queueing model with catastrophes and geometric reneging, the time-dependent probability was established using PGF, further explicit expressions and computational schemes for various system characteristics were derived. Recently, many researchers (c.f. [20–24]) have been working on the subject.

The standby switching failure concept in the machine repair problem (MRP) was first introduced by Lewis [25]. Then, Ke et al. [26] studied the machine repair problem with standby switching failure and general repair times. They used the supplementary variable method to find the steady-state probability. Recently, Shekhar et al. [27] focused on the MRP model with common cause failure, reboot delay, and switching failure. The steady-state probabilities were found via the successive overrelaxation (SoR) method. For recent works on switching failure, the reader may refer to ([28–31]). In this investigation, we dealt with impatience behavior in the waiting line, which is effectively prevailing the service system. The machines continue to break down and the failed machines may renege in the synchronized pattern due to the long wait experienced at a queue whenever the repairman is absent on working vacation.

The chapter is organized as follows. In Section 10.3, we describe the model and present Chapman–Kolmogorov equations. In Sections 10.4 and 10.5, we present the steady-state probability of the system, and derive different system characteristics, as well as the expected total cost. In Section 10.6, we present various numerical results that illustrate the effect of the various parameters on the expected total cost of the model. We conclude the chapter in Section 10.7, where we discuss possible generalizations and extensions.

10.3 MODEL DESCRIPTION AND STATE PROBABILITIES

Consider a multi-components machining system having M operating units with the provisioning of S warm spare units in addition to working vacation and vacation interruption of the server, switching failure of the spare unit, and discouragement behavior of the caretaker of failed machines. In order to formulate the model mathematically, the following assumptions are made:

- The system will work with at least m operating units in short mode where for normal functioning M units are required. The system capacity is finite, say L.
- The lifetime of operating units and spare units are assumed to be exponentially distributed with parameter λ and ν ($< \lambda$)
- On the failure of operating unit, the available warm spare unit switches in negligible time in place of a failed operating unit. The switching process may be efficacious in geometric fashion with switching failure probability, q. The process continues until all spare units fail to switch.
- On the exhaustion of all warm spare units, the lifetime of operating units diminishes due to overload in short mode. In short mode, the failure of operating units follows Poisson process with degradable parameter $\lambda_d (> \lambda)$
- The state-of-the-art repair facility consists of a single permanent repairman who seeks for working vacation to maintain economical repair facility. On the completion of repair of all failed units in the system, repairman opts for multiple working vacations of the random period.
- The vacation time for the repairman follow an exponential distribution with mean time $1/\theta$. After returning from vacation, if the repairman finds no failed machine in the system, it goes on working vacation.
- The repair time of failed units follows an exponential distribution with parameter μ_b *and* μ_v in busy mode and working vacation mode respectively.
- The technician repairs the failed units in First Come First Served (FCFS) fashion. After the repair, the unit joins the pool of operating units, if the system is at a lower rate; otherwise, the unit joins a group of warm spare units.
- For just-in-time repair facility in short mode, vacation is interrupted immediately. On the completion of the repair of any failed machines in short mode, the technician starts to serve in normal busy mode.
- If the server is in working vacation mode, the caretaker of the failed machines may turn out to be impatient and may perform synchronized abandonments from the system in the following manner: an extraneous repair facility is set on and it arrives at the system according to a Poisson process at rate ζ. Every arrival epoch of the extraneous facility constitutes an abandonment opportunity for the present caretaker of failed machines. We suppose that each one of them decides to abandon the system with probability p or remains in the system with complementary probability $1 - p$ independently of the others.
- All events are statistically independent of each other.

10.4 CHAPMAN–KOLMOGOROV EQUATION

The states of the machining system are structured in notation by the pair

$$\{(I(t),N(t))|N(t)=0,1,\ldots,L;t\geq 0\},$$

Where,

$I(t)=0$, Repairman is in a working vacation at time t,
$I(t)=1$, Repairman is in a working vacation at time t.

The transient-state probabilities at time t (≥ 0)are defined as follows:

$$P_{0,n}(t)=Prob\{I(t)=0,N(t)=n\};n=0,1,2\ldots L,$$

$$P_{1,n}(t)=Prob\{I(t)=1,N(t)=n\};n=1,2,\ldots,L.$$

As $\rightarrow\infty$, the system tends to be stable. The governing Chapman–Kolmogorov equations are as follows:

Case I: When repairman is on working vacation

$$-(M\lambda+Sv)P_{0,0}+\mu_v P_{0,1}+\xi\left\{\sum_{i=1}^{L}p^i P_{0,i}\right\}+\mu_b P_{1,1}=0, \tag{10.1}$$

$$-(M\lambda+(S-1))v+\mu_v+\xi+\theta)P_{0,1}+(M\lambda(1-q)+Sv)P_{0,0}+$$
$$\mu_v P_{0,2}+\xi\left\{\sum_{i=1}^{L}\binom{i}{i-1}p^{i-1}(1-p)P_{0,i}\right\}=0, \tag{10.2}$$

$$-(M\lambda+(S-1))v+\mu_v+\xi+\theta)P_{0,n}+(M\lambda(1-q)+(S-n+1)v)P_{0,n-1}+$$
$$\sum_{i=0}^{n-2}M\lambda(1-q)q^{n-i-1}P_{0,i}+\mu_v P_{0,n+1}+\xi\left\{\sum_{i=n}^{L}\binom{i}{i-n}p^{i-n}(1-p)^n P_{0,i}\right\}$$
$$=0,\ 2\leq n\leq S-1, \tag{10.3}$$

$$-\left(M\lambda_d+\mu_v+\xi+\theta\right)P_{0,S}+\left(M\lambda\left(1-q\right)+v\right)P_{0,S-1}+$$
$$\sum_{i=0}^{S-2}M\lambda\left(1-q\right)q^{S-i-1}P_{0,i}+\xi\left\{\sum_{i=S}^{L}\binom{i}{i-S}p^{i-S}\left(1-p\right)^{S}P_{0,i}\right\}=0, \tag{10.4}$$

$$-\left(\left(M-1\right)\lambda_d+\mu_v+\xi+\theta\right))P_{0,S+1}+M\lambda_dP_{0,s}+\sum_{i=0}^{S-1}M\lambda\, q^{S-i}P_{0,i}+$$
$$\xi\left\{\sum_{i=S+1}^{L}\binom{i}{i-S-1}p^{i-S-1}\left(1-p\right)^{S+1}P_{0,i}\right\}=0, \tag{10.5}$$

$$-\left(\left(M+S-n\right)\lambda_d+\mu_v+\xi+\theta\right)P_{0,n}+\left(M+S-n+1\right)\lambda_dP_{0,n-1}+$$
$$\xi\left\{\sum_{i=n}^{L}\binom{i}{i-n}p^{i-n}\left(1-p\right)^{n}P_{0,i}\right\};S+2\le n\le L-1=0, \tag{10.6}$$

$$-\left(\mu_v+\xi+\theta\right)P_{0,L}+m\lambda_dP_{0,L-1}+q^{L}\xi\,P_{0,L}=0. \tag{10.7}$$

Case II: When the repairman is in the busy state

$$-\left(M\lambda+\left(S-1\right)v+\mu_b\right)P_{11}+\mu_bP_{12}+\theta P_{01}=0, \tag{10.8}$$

$$-\left(M\lambda+\left(S-1\right)v+\mu_b\right)P_{1,n}+\left(M\lambda\left(1-q\right)+\left(S-n+1\right)v\right)P_{1,n-1}+$$
$$\mu_bP_{1,n+1}+\sum_{i=1}^{n-2}M\lambda\left(1-q\right)q^{n-i-1}P_{1,i}+\theta P_{0,n}=0;2\le n\le S-1, \tag{10.9}$$

$$-\left(M\lambda_d+\mu_b\right)P_{1,S}+\left(M\lambda\left(1-q\right)+v\right)P_{1,S-1}+\mu_bP_{1,S+1}+$$
$$\sum_{i=1}^{S-2}M\lambda\left(1-q\right)q^{S-i-1}P_{1,i}+\theta P_{0,S}+\mu_vP_{0,S+1}=0, \tag{10.10}$$

$$-\left(\left(M-1\right)\lambda_d+\mu_b\right)P_{1,S+1}+M\lambda_dP_{1,S}+\mu_bP_{1,S+2}+$$
$$\sum_{i=1}^{S-1}M\lambda q^{S-i}P_{1,i}+\theta P_{0,S+1}+\mu_vP_{0,S+2}=0, \tag{10.11}$$

$$-\left(\left(M+S-n\right)\lambda_d+\mu_b\right)P_{1,n}+\left(M+S-n+1\right)\lambda_d P_{1,n-1}+$$
$$\mu_b P_{1,n+1}+\theta P_{0,n}+\mu_v P_{0,n+1}=0, S+2\le n\le L-1, \tag{10.12}$$

$$-\mu_b P_{1,L}+m\lambda_d P_{1,L-1}+\theta P_{0,L}=0. \tag{10.13}$$

For above-defined state probabilities, the normalizing conditions

$$\sum_{n=0}^{L} P_{0,n}+\sum_{n=1}^{L} P_{1,n}=1. \tag{10.14}$$

10.5 THE STEADY-STATE SOLUTION

The system of simultaneous linear Equations (10.1)–(10.13) is converted to matrix form as

$$\boldsymbol{AP}=\boldsymbol{0} \tag{10.15}$$

Here, $\boldsymbol{A}$ is a (L+1) square matrix, coefficient of state probabilities $\boldsymbol{P}_{0,i}$ and $\boldsymbol{P}_{1,i}$ are the element of $\boldsymbol{A}$. $\boldsymbol{P}$ is a column vector of unknown state probabilities of order $(L+1)\times 1$ and $\boldsymbol{0}$ is a null vector of suitable dimension. By the normalizing condition given in Equation (10.14), we have

$$\boldsymbol{Pe}=\boldsymbol{1}, \tag{10.16}$$

where $\boldsymbol{e}$ is a unit vector. The above system of linear Equation (10.15) can be represented in

$$\boldsymbol{CP}=\boldsymbol{B}, \tag{10.17}$$

where $\boldsymbol{C}$ is the same matrix as $\boldsymbol{A}$ except each element in the last row is replaced by 1 and $\boldsymbol{B}$ is a column vector having zero elements except the last element which is replaced by 1.

We solve the non-homogeneous system of linear Equation (10.17) represented in a matrix form, and obtain the stationary probabilities using the G.E.E.(Gauss elimination extended) numerical technique SOR (successive over-relaxation) method with over-relaxation parameter value 1.25 in MATLAB® (R2019b) software.

In the next section, the system characteristics of the queueing system under consideration are established in terms of the state probabilities.

10.6 SYSTEM PERFORMANCE MEASURES

The performance indices of the machining system can be used not only to judge the efficiency and behavior of the existing system, but also for future design and development of the same.

- The expected number of failed machines in the system

$$E_N = E(N) = \sum_{i=0}^{1} \sum_{n=i}^{L} nP_{i,n} \tag{10.18}$$

- The throughput of the system

$$TP = \sum_{n=1}^{L} \mu_v P_{0,n} + \sum_{n=1}^{L} \mu_b P_{1,n} \tag{10.19}$$

- Mean number of standby machines in the system

$$E_S = \sum_{i=0}^{1} \sum_{n=i}^{S} (S-n) P_{i,n} \tag{10.20}$$

- Mean number of operating machines in the system

$$E_0 = M \sum_{i=0}^{1} \sum_{n=i}^{S} P_{i,n} + \sum_{i=0}^{1} \sum_{n=S+1}^{L} (M+S-n) P_{i,n} \tag{10.21}$$

- Expected carrying load of failed machines in the system

$$\lambda_{eff} = M \sum_{i=0}^{1} \sum_{n=i}^{S} \left(M\lambda + (S-n)v\right) P_{i,n} + \sum_{i=0}^{1} \sum_{n=S+1}^{L-1} (M+S-n) \lambda_d P_{i,n} \tag{10.22}$$

- Expected waiting time of the failed operating machine in the system

$$E_W = \frac{E_N}{\lambda_{eff}} \tag{10.23}$$

- Expected delay time in the service by the repairman

$$E_D = \frac{E_N}{TP} \tag{10.24}$$

- Effective switching failure rate of the standby machine

$$SR = \sum_{i=0}^{1} \sum_{n=i}^{S-1} M\lambda q P_{i,n} \tag{10.25}$$

- Effective reneging rate during the vacation period

$$RR = \sum_{n=i}^{L} \left(1-p^n\right) \xi P_{0,n} \tag{10.26}$$

- Failure frequency of the system

$$FF = m\lambda_d \sum_{n=i}^{1} P_{i,L} \tag{10.27}$$

- Vacation interruption frequency of the system

$$VF = \sum_{n=S+1}^{L} \mu_v P_{0,n} \tag{10.28}$$

- Machine availability in the system

$$MA = 1 - \frac{E_N}{M+S}. \tag{10.29}$$

10.6.1 Expected Total Cost

To have a quantitative idea about the system performance, we formulate a cost function using various cost element associated with performance measures. Let us define the cost elements as follows:

C_H	fixed cost per unit time of failed machine waiting in the system,
C_S	fixed cost per unit time of failed machine waiting in the system,
C_0	fixed cost per unit time of available operating machine in the system,
C_F	fixed switching failure cost per unit time,
C_R	fixed cost per unit time associated with reneging of failed machine,
C_{mb}	Service cost per unit time per customer during a normal working period μ_b
C_{mv}	Service cost per unit time per customer during a working vacation period μ_v
C_{th}	fixed cost per unit time for server being in working vacation,
C_X	fixed cost per unit time for extraneous service facility.

Employing the definitions of each cost element listed above and its corresponding system characteristics, the expected total cost function per unit time is framed as

$$\text{TC} = \text{C}_\text{H}\text{E}_\text{N} + \text{C}_\text{S}\text{E}_\text{S} + \text{C}_0\text{E}_0 + \text{C}_\text{F}\text{SR} + \text{C}_\text{R}\text{RR} + \text{C}_\text{th}\theta + \text{C}_\text{X}\xi + \text{C}_\text{mb}\mu_\text{b} + \text{C}_\text{mn}\mu_\text{v}\cdot$$

10.6.2 The Quasi-Newton Method

The essence of the quasi-Newton method is to find a search direction in each iteration. We use different steps along this direction for a better solution until the tolerance is small enough. Define the vector $\vec{\Omega} = \left[\mu_v, \mu_b\right]^T$ and the respective gradient vector $\vec{\nabla} TC\left(\vec{\Omega}\right)$ which consists of $\frac{\partial TC}{\partial \mu_v}$ and $\frac{\partial TC}{\partial \mu_b}$. Next, we employ the quasi-Newton method to find the global minimum expected cost and corresponding decision variables by using the following steps:

Step 1: Let $\overrightarrow{\Omega_0} = \left[\mu_v, \mu_b\right]^T$
Step 2: Set the initial trial solution for $\vec{\Omega}_0$ and compute $TC\left(\overrightarrow{\Omega_0}\right)$

Step 3: Compute the cost gradient $\vec{\nabla}TC\left(\vec{\Omega}\right)=\left[\frac{\partial TC}{\partial \mu_v},\frac{\partial TC}{\partial \mu_b}\right]^T\Bigg|_{\vec{\Omega}_0}$ and the cost Hessian Matrix

$$H\left(\vec{\Omega}\right)=\begin{pmatrix} \frac{\partial^2 TC}{\partial \mu_v^{\ 2}} & \frac{\partial^2 TC}{\partial \mu_v \partial \mu_b} \\ \frac{\partial^2 TC}{\partial \mu_v \partial \mu_b} & \frac{\partial^2 TC}{\partial \mu_b^{\ 2}} \end{pmatrix}$$

Step 4: Find the new trial solution $\overrightarrow{\Omega_{n+1}}=\overrightarrow{\Omega_n}-\left[H\left(\overrightarrow{\Omega_n}\right)\right]^{-1}\vec{\nabla}TC\left(\overrightarrow{\Omega_n}\right)$

Step 5: Set $n = n + 1$ and repeat steps 2 – 4 until

$$\left|\frac{\partial TC}{\partial \mu_v}\right| < \varepsilon_1 \text{ and } \left|\frac{\partial TC}{\partial \mu_b}\right| < \varepsilon_2,$$

where $\varepsilon_1 = \varepsilon_2 = 10^{-7}$ are the tolerances.

Step 6: Find the global minimum value $TC\left(\overrightarrow{\Omega_n^T}\right)=TC\left(\mu_v,\mu_b\right)$.

10.7 NUMERICAL RESULTS

The analytical results of the machine repairable system's performance measures are not sufficient to establish the virtue of the developed model. To scrutinize the practical applicability of the proposed repairable Markovian model with vacation interruption, synchronized and switching failure, several numerical experiments are liquidate in MATLAB (R2019b) software and the results are presented in Tables 10.1 to 10.2

TABLE 10.1
The Optimal Services Rates and Expected Total Cost for Different System Parameters

(M, S, λ, λ_d, ν)	μ_v^*	μ_b^*	*TC∗*	**Total Iterations**
(8, 4, 0.1, 0.5, .04)	4.3727	6.8809	362.85	11
(10, 4, 0.1, 0.5, .04)	5.5318	8.427	419.66	10
(12, 4, 0.1, 0.5, 0.04)	6.6802	9.9814	476.4	13
(14, 5, 0.1, 0.5, 0.04)	7.4629	10.918	551.58	11
14, 6, 0.1, 0.5, 0.04	7.2508	10.371	571.71	11
14, 7, 0.1, 0.5, 0.04	7.1268	9.8987	593.04	13
14, 8, 0.2, 0.5, 0.04	10.773	12.596	669.58	12
14, 8, 0.3, 0.5, 0.04	13.126	15.239	715.85	12
14, 8, 0.4, 0.5, 0.04	14.341	17.524	756.99	12
8, 4, 0.1, 0.6, 0.04	4.5427	7.4509	366.58	13
(8, 4, 0.1, 0.7, 0.04)	4.6912	8.0066	370.21	14
(8, 4, 0.1, 0.8, 0.04)	4.8221	8.5488	373.74	12
(8, 4, 0.1, 0.5, 0.06)	4.4597	6.9041	363.49	11
(8, 4, 0.1, 0.5, 0.07)	4.5025	6.9157	363.8	11
(8, 4, 0.1, 0.5, 0.08)	4.5449	6.9275	364.12	11

TABLE 10.2
The Optimal Services Rates and Expected Total Cost for Different System Parameters

(M,S,P,q,θ)	μ_v^*	μ_b^*	TC^*	Total Iterations
(8, 4, 0.1, 0.5, 1)	5.6068	6.6317	354.38	11
(10, 4, 0.1, 0.5, 1)	6.7674	8.1803	411.25	10
(12, 4, 0.1, 0.5, 1)	7.9176	9.7371	468.04	13
(10, 5, 0.1, 0.5, 1)	6.7565	7.7018	432.33	11
(10, 6, 0.1, 0.5, 1)	6.8542	7.2775	454.44	13
(10, 7, 0.1, 0.5, 1)	6.9205	6.9206	477.28	19
(8, 4, 0.2, 0.5, 1)	5.5583	6.6269	354.17	11
(8, 4, 0.3, 0.5, 1)	5.5105	6.6234	353.98	11
(8, 4, 0.4, 0.5, 1)	5.4645	6.6212	353.81	13
(12, 4, 0.1, 0.6, 1)	8.2675	10.674	482.86	13
(12, 4, 0.1, 0.7, 1)	8.6247	11.596	498.49	12
(12, 4, 0.1, 0.8, 1)	9.0072	12.482	514.85	14
(10, 5, 0.1, 0.5, 2)	5.4503	8.0126	441.27	10
(10, 5, 0.1, 0.5, 3)	4.1456	8.2527	449.69	10
(10, 5, 0.1, 0.5, 4)	2.8665	8.4449	457.77	9

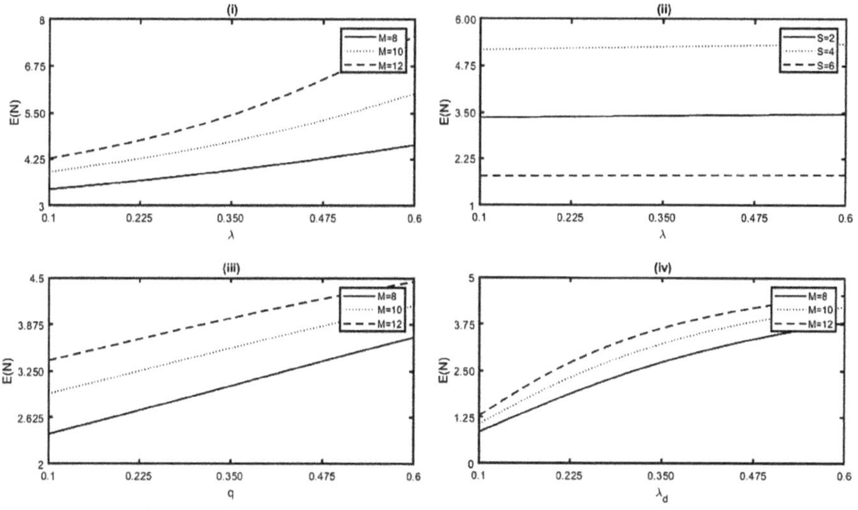

FIGURE 10.1 Total cost for varying different parameters.

and Figures 10.1 & 10.2. For this purpose $M = 8$; $S = 4$; $\lambda = 0.1$; $\lambda_d = 0.5$; $v = 0.08$; $\xi = 0.2$; $\theta = 2$; $p = 0.4$; $q = 0.5$; $m = 3$; Figure 10.1 depicts the variability of the expected number of failed machine $E(N)$ with respect to the system parameters. If we increase the failure rate as (λ, ν) then we found that $E(N)$ also increases, as intuitively expected. For Tables 10.1 to 10.2, we take $M = 10$; $S = 5$; $p = 0.1$; $q = 0.5$; $\theta = 4$; $\xi = 0.2$; $m = 3$; $\lambda = 0.1$; $\lambda_d = 0.5$; $v = 0.04$; $CF = 50$; $CS = 25$; $CO = 20$; $CSW = 17$; $CR = 19$; $CT = 10$; $cv = 3$; $cb = 5$; with $\mu_v = 1$ and $\mu_b = 2$; for iteration.

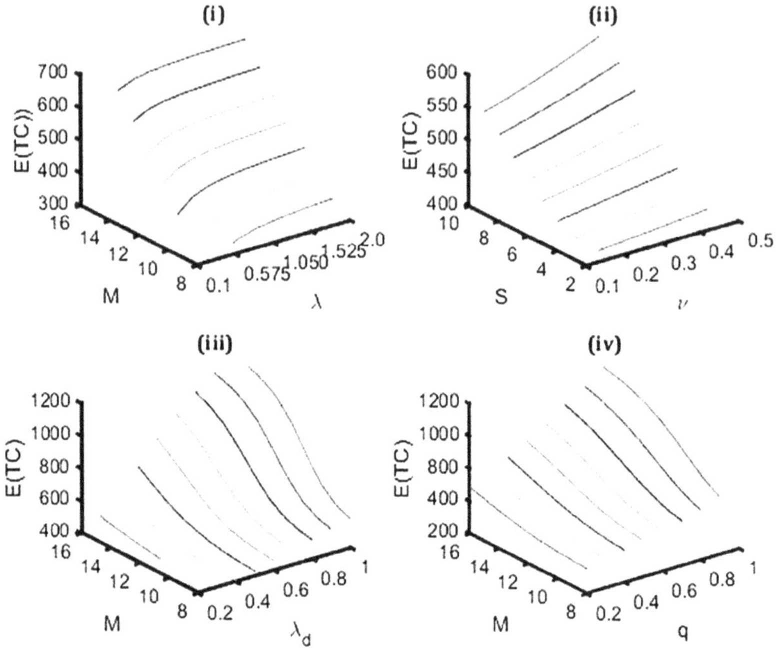

FIGURE 10.2 Total cost for varying different parameters.

The numerical results of the joint optimal values $\left(\mu_v^*, \mu_b^*\right)$ and the expected total cost *(E(TC))* are presented in Tables 10.1 and 10.2. It is well observed that by increasing the standby machine, the expected cost significantly increases. Further, if the components are more prone to failure at a higher rate, a large service rate is required to accomplish optimum service conditions. It is an obvious result and supports our methodology and analysis. A similar kind of impact is also tabulated in Table 10.1 for system parameters (λ_d, λ, v) as (λ_d, λ) are increased, then the optimal service rate and expected total cost also increase.

Tables 10.1 and 10.2 comparisons the results of optimal repair rate/service rate $\left(\mu_v^*, \mu_b^*\right)$ and corresponding expected total cost (E(TC)) if we increase the standby machine then the expected cost also increase. We observe that if components are more prone to failure at a higher rate, a better service rate is required to accomplish optimum service conditions. It is an obvious result and supports our methodology and analysis. A similar kind of impact is also tabulated in Table 10.1 for system parameters (λ_d, λ, v). As (λ_b, λ) increased, then the optimal service rate and expected total cost also increase.

In Table 10.2, as *p* increases, optimal repair rate and expected total cost decrease. In addition, if switching failure probability (q) and vacation rate θ increase, the optimal service rate and expected total cost also increase.

10.8 CONCLUSION

In this chapter, we gave an analysis for the single server machining system with switching failure and synchronized reneging. Some real-life phenomena, like a working vacation, vacation interruption of the server, and discouragement behavior of the caretaker of failed machines have also been incorporated. The governing Chapman-Kolmogorov equations have been formulated and solved by employing the SOR method to obtain the steady-state probabilities. Various essential performance characteristics for comparative and optimal analysis have been developed. Finally, numerical experiments were given to demonstrate how the various system parameters affect the joint optimal values $\left(\mu_v^*, \mu_b^*\right)$ and the expected total cost *(E(TC))*. This model can be used for applications in various machining systems as power generation, telecommunications networks, transmission systems, relay stations, etc.

The present study may be extended with *T*-policy, *N*-policy, *D*-policy, common cause failure, setup time, and impatience behavior of caretaker as reneging and balking. We can look forward to some Ant colony optimization and particle swarm optimization (PSO).

REFERENCES

[1] Wang, K. H., Ke, J. B., & Ke, J. C. (2007). Profit analysis of the M/M/R machine repair problem with balking, reneging, and standby switching failures. *Computers & Operations Research*, *34*(3), 835–847.

[2] Jain, M., & Sanga, S. S. (2017). Control F-policy for fault tolerance machining system with general retrial attempts. *National Academy Science Letters*, *40*(5), 359–364.

[3] Miyagi, P. E., & Riascos, L. A. M. (2006). Modeling and analysis of fault-tolerant systems for machining operations based on Petri nets. *Control Engineering Practice*, *14*(4), 397–408.

[4] Malik, S., & Huet, F. (2011). *Adaptive fault tolerance in real time cloud computing*. In *2011 IEEE World Congress on Services*, 280–287. IEEE.

[5] Shekhar, C., Kumar, A., Varshney, S., & Ammar, S. I. (2019). Fault-tolerant redundant repairable system with different failures and delays. *Engineering Computations*, *37*(3), 1043–1071

[6] Jain, M., & Meena, R. K. (2020). Availability analysis and cost optimization of M/G/1 fault-tolerant machining system with imperfect fault coverage. *Arabian Journal for Science and Engineering*, *45*(3), 2281–2295.

[7] Hsu, Y. L., Ke, J. C., Liu, T. H., & Wu, C. H. (2014). Modeling of multi-server repair problem with switching failure and reboot delay and related profit analysis. *Computers & Industrial Engineering*, *69*, 21–28.

[8] Servi, L. D., & Finn, S. G. (2002). M/M/1 queues with working vacations (M/M/1/WV). *Performance Evaluation*, *50*(1), 41–52.

[9] Laxmi, P. V., & Rajesh, P. (2020). Performance characteristics of discrete-time queue with variant working vacations. *International Journal of Operations Research and Information Systems (IJORIS)*, *11*(2), 1–21.

[10] Jain, M., Kumar, P., & Sanga, S. S. (2020). Fuzzy Markovian modeling of machining system with imperfect coverage, spare provisioning and reboot. *Journal of Ambient Intelligence and Humanized Computing*, *12*, 1–13.

[11] Patterson, J., & Korzeniowski, A. (2020). Decomposition of M/M/1 with Unreliable Service and a Working Vacation. *International Journal of Statistics and Probability*, *9*(1), 1–63.

[12] Kumar, A. (2020). Single server multiple vacation queue with discouragement solve by confluent hypergeometric function. *Journal of Ambient Intelligence and Humanized Computing*, 1–12.

[13] Xu, X., & Wang, H. (2019). Analysis of fluid model modulated by an M/PH/1 working vacation queue. *Journal of Systems Science and Systems Engineering*, *28*(2), 132–140.
[14] Wang, H., Xu, X., & Wang, S. (2019). Equilibrium customers strategies in the Markovian working vacation queue with setup times. *International Journal of Computing Science and Mathematics*, *10*(5), 443–458.
[15] Sethi, R., & Bhagat, A. (2019). *Performance analysis of machine repair problem with working vacation and service interruptions*. In *AIP Conference Proceedings*, 2061(1), Publishing LLC.
[16] Yang, D. Y., & Wu, C. H. (2019). Performance analysis and optimization of a retrial queue with working vacations and starting failures. *Mathematical and Computer Modelling of Dynamical Systems*, *25*(5), 463–481.
[17] Shekhar, C., Varshney, S., & Kumar, A. (2020). Reliability and vacation: The critical issue. In Ram, M. & Phan, H. (eds) *Advances in Reliability Analysis and its Applications*, 251–292. Springer, Cham.
[18] Adan, I., Economou, A., & Kapodistria, S. (2009). Synchronized reneging in queueing systems with vacations. *Queueing Systems*, *62*(1), 1–33.
[19] Dimou, S., & Economou, A. (2013). The single server queue with catastrophes and geometric reneging. *Methodology and Computing in Applied Probability*, *15*(3), 595–621.
[20] Panda, G., & Goswami, V. (2020). Strategic customers in Markovian queues with vacations and synchronized abandonment. *The ANZIAM Journal*, *62*(1), 89–120.
[21] Kumar, R. (2016). A single-server Markovian queuing system with discouraged arrivals and retention of reneged customers. *Yugoslav Journal of Operations Research*, *24*(1).
[22] Hou, J., & Knaze, J. (2019). The effect of exchange rate regimes on business cycle synchronization: A robust analysis.
[23] Sun, W., Li, S., & Tian, N. (2020). Comparisons of customer behavior in Markovian queues with vacation policies and geometric abandonments. *RAIRO-Operations Research*, *54*(3), 615–636.
[24] Panda, G., Goswami, V., & Banik, A. D. (2016). Equilibrium and socially optimal balking strategies in Markovian queues with vacations and sequential abandonment. *Asia-Pacific Journal of Operational Research*, *33*(05), 1650036.
[25] Boehm, F., & Lewis, E. E. (1992). A stress-strength interference approach to reliability analysis of ceramics: Part I—fast fracture. *Probabilistic Engineering Mechanics*, *7*(1), 1–8.
[26] Ke, J. C., Liu, T. H., & Yang, D. Y. (2016). Machine repairing systems with standby switching failure. *Computers & Industrial Engineering*, *99*, 223–228.
[27] Shekhar, C., Kumar, A., & Varshney, S. (2020). Load sharing redundant repairable systems with switching and reboot delay. *Reliability Engineering & System Safety*, *193*, 106656.
[28] Shekhar, C., Raina, A., Kumar, A. & Iqbal, J. (2017). A survey on queues in machining system: Progress from 2010 to 2017, *Yugoslav Journal of Operations Research*, *27*(4). 391–413.
[29] Ma, J., Chai, Z., Zhang, W., Govoreanu, B., Zhang, J. F., Ji, Z., & Jurczak, M. (2016). *Identify the critical regions and switching/failure mechanisms in non-filamentary RRAM (a-VMCO) by RTN and CVS techniques for memory window improvement*. In *2016 IEEE International Electron Devices Meeting (IEDM)*, 21–4. IEEE.
[30] Shen, J., Hu, J., & Ye, Z. S. (2020). Optimal switching policy for warm standby systems subjected to standby failure mode. *IISE Transactions*, *52*(11), 1262–1274.
[31] Wu, C. H., & Yang, D. Y. (2020). Dynamic control of a machine repair problem with switching failure and unreliable repairmen. *Arabian Journal for Science and Engineering*, *45*(3), 2219–2234.

11 Optimal Cluster Head Election in Industrial WSNs Using the Multi-Objective Genetic Algorithm

R. Pal, A. K. Malik, S. Yadav, and R. Karnwal

CONTENTS

11.1 INTRODUCTION

Sensor networks are expected to consist of various sensor nodes and one or many base stations or sinks. A sensor node is a small autonomous device consisting of several constraints, including the battery capacity, memory, and transmission range [1]. All sensor nodes have the transceivers to gather the data from the surrounding environment and transfer it to the sink, where the sensed information is stored, processed and made available for the end user [2].

Recently, industrial wireless sensor networks (IWSN) have emerged as the prominent technology in today's industrial marketplace. It provides low-cost solutions for efficient production and maintenance in the industrial environment. The wireless networks of thousands of tiny and smart motes are used to make the industries self-healing and reliable [3]. These networks are far better than the wired networks used in the industries, due to various reasons such as mobility support, can be deployed in harsh locations. There is no need of cables, and they are dynamic in nature. The three key applications areas of the IWSN are: process automation, condition monitoring, and phenomena sensing [3]. However, the sensor nodes are battery operated or there

DOI: 10.1201/9781003089636-11

are some restrictions on the power supply availability. Sometimes these sensors are deployed randomly and supposed to perform the action properly and effectively. Results of this cause the varying degree of node density in the area [4]. Some segments of the network have high node density while others may have a smaller number of nodes which make the network sparse. Because of the energy constraints, the nodes in the sparse region will soon die and this segment will become unattended. Generally, sensor networks are deployed in harsh environments and, because of the environmental conditions, some sensors nodes will become inoperable or faulty. Therefore, it is desirable that these networks should be fault-tolerant [5]. However, placing the new sensor nodes instead the terminated ones is not the desirable solution to this problem. The topology of the sensor network is changing regularly. Therefore, clustering is one of the most promising solutions for organizing the sensor network [6]. In order to implement clustering efficiently and make clusters which consume less energy in data transmission, various algorithms have been proposed in the literature [7–10]. Among the most renowned of these classical protocols are LEACH [11], SEP [12], MSEP-E [13], FSEP-E [8], PEGASIS [14], HEED [15] and many more. These algorithms group various sensor nodes and select one leader for each formed cluster, which is known as the cluster head (CH). The sensed data from each node is collected by the respective CH of the cluster and all CHs send the data to the sink node.

In order to satisfy coverage requirements, the sensor is deployed densely so that some of the nodes are in sleep mode to save energy. There are various methods proposed in the literature for CH selection. However, selecting an optimal number of CHs is very challenging as we require various combination of different nodes to be elected as the CH. Therefore, clustering can be mapped as a NP-Hard problem [16]. Recently, various meta-heuristic algorithms have been used to solve clustering as an optimization problem [17–20]. Many real-world problems in terms of achieving several objectives such as minimizing cost, minimizing risk and maximizing reliability, and many more are considered to be a single objective optimization problem [25–30]. The main aim is to maximize or minimize the value of a single objective function, which is the result of lumping many objectives into one [6]. This cannot provide a bundle of alternative solutions which can deal with multiple objectives. On the other hand, in a multi-objective optimization, a set of compromised solution has been generated by interacting the considered objectives, widely known as the trade-off, non-dominated solutions or Pareto-optimal solutions [21].

The network lifetime of the sensor network is largely affected by the selection of CHs. The CH having high node density, high remaining node energy and the least node to sink distance is most eligible for the selection. Various evolutionary methods have used for clustering the sensor networks, including HCR (hierarchical clustering algorithm) [22] and ERP (evolutionary routing problem) [23], which used cluster compactness, energy, separation and other factors to formulate the fitness function. However, some of the considered factors in single objective fitness function are contradictory in nature. Therefore, it is better to consider separate objectives for each considered factor. Hence, in the proposed method, the multi-objective genetic algorithm [24] is considered to find the relevant CHs in the sensor network.

Further, the chapter is organized as follows: Section 11.2 given the basic idea of MOGA algorithm followed by Section 11.3, which explains the proposed method. The results are explained and discussed in Section 11.4, followed by the conclusion in Section 11.5.

11.2 MULTI-OBJECTIVE GENETIC ALGORITHM (MOGA)

A Multi-Objective Genetic Algorithm (MOGA) [24] is a multi-objective version of genetic algorithm. The method of assigning fitness distinguishes it from GA. The main steps of MOGA are discussed below:

- Check for the dominance of each solution in the population. Rank of a solution can be given by Equation (11.1).

$$r_i = 1 + n_i \tag{11.1}$$

Where, i represents i^{th} solution, r_i denotes of rank of solution i, and n_i is the number of solutions that dominates solution i. If i is a non-dominated solution (i.e., number of solutions that dominate i is zero (n_i=0)), then the rank of solution i will be 1. Moreover, the maximum rank of a solution can be N, where N is the total number of solutions. All solutions calculate their rank values based on domination levels.

- The second main aim of MOGA is to maintain the diverse set of solutions. By niching among each solution of each rank, anyone can make the solution diverse. The niche count denotes the density of each solution i and is given by Equation (11.2).

$$nci = \sum_{j=1}^{\mu(r_i)} sh(dij) \tag{11.2}$$

Where, number of solutions of rank r_i is given by $\mu(r_i)$ and $sh(d_{ij})$ is the sharing function defined over distance d and given by Equation (11.3).

$$sh(d) - \begin{cases} 1 - \left(\dfrac{d}{\sigma \ share}\right)^{\alpha} & \text{if } d \leq \sigma \ share \\ 0 & \text{otherwise} \end{cases} \tag{11.3}$$

Here, d is the distance between any two solutions and $\alpha = 1$. With the help of objective function values, the distance metric can be computed. Further, Equation (11.4) gives the normalized distance based on maximum fitness (f_k^{max}) and minimum fitness (f_k^{min}) in the kth objective, from a solution i to a solution j.

$$dij = \sqrt{\sum_{k=1}^{M} \frac{f_k(i) - f_k(j)}{f_k^{\max} - f_k^{\min}}} \tag{11.4}$$

The distance will be calculated for each solution i and j.

- The summation of sharing function values gives the niche count. Now, shared fitness value for each solution can be found by dividing the fitness (F) by niche count (nc). Therefore, each solution having the same rank will have the similar fitness. After all this, the calculation will shift to the next rank solution and identical process will be performed.
- Choosing the maximum rank r^* in r_i which has $r_i > 0$. By sorting fitness by rank and fitness averaging yields the average fitness to any solution. The shared and scaled shared fitness value of solution j is given by Equation (11.5) and Equation (11.6) respectively.

$$F_j' = \frac{F_j}{nc_j} \tag{11.5}$$

$$F_j' = \frac{F_j \mu(r)}{\sum_{k=1}^{\mu(r)} F_j'} \tag{11.6}$$

- Repeat the above process till $r < r*$.

11.3 MOGA-BASED CH SELECTION

In this chapter, a new clustering approach is proposed based on MOGA. It can be divided into two phases: the first one is the cluster set-up phase and the second can be the steady-state phase. The detailed description of these two phases are explained below.

11.3.1 Set-up Phase

This consists three main phases: cluster head advertisement, cluster formation and creation of transmission schedule. Firstly, CH sends an advertisement packet to the cluster nodes to inform that they have become cluster head based on MOGA. In the second step cluster nodes receives an advertisement from the cluster heads and then cluster nodes send the request to join cluster and become the cluster members under that cluster head. Moreover, all non-cluster head nodes turn off their transmitter to save the transmission energy and ON when then have something to transmit. Now in the third step each cluster head schedules a transmission slot for their cluster nodes. Each node follows its allocated schedule to transmit the data.

1) *Cluster Heads selection using MOGA:* For performing clustering using MOGA, first calculate the scaled fitness by the above-described procedure. Scaled fitness will perform the main role in clustering as having greater fitness value have more chance to become a cluster head. The cluster with the greatest energy will die later. Here two objective functions are used in the MOGA. Definitions of the objective function are given below:
Objective 1: Minimize the compactness
Objective 2: Maximize the separation.

11.3.2 Steady-State Phase

In this phase, the cluster node send data to the cluster head (CH) nodes. A single hop transmission is only followed to communicate between the member sensor of cluster and CH. Now aggregated data from the cluster head are forwarded to the base station either directly or by other cluster heads.

11.4 EXPERIMENT RESULTS AND DISCUSSION

The simulation of the small industrial WSN is performed in MATLAB® in the phenomena having area 200×100 m^2. The nodes in the area are randomly deployed. There are two types of nodes, the normal node and the advanced node. The battery capacity of the normal nodes is 0.5 Joules whereas the advanced nodes have a battery of 1 Joule capacity. The two different nodes are used to maintain the heterogeneity in the network. There are two scenarios of network, one type of network is having only 10% advanced nodes in the network while in other type 20% advanced node are used. The sink is positioned outside the sensor network at position (200, 50). A snapshot of the network when performing the simulation is depicted in Figure 11.1. In the figure, * represents the cluster head, '+' denotes the advanced nodes and '△' denotes the normal nodes. The sink, represented by the filled blue circle, is placed outside the sensor region. Further, when any of the nodes in the network depletes its energy completely, its color is converted to red; when all of the nodes have died the entire network has nodes with red color only. Figure 11.2 shows the snapshot of the same scenario.

The clustering of sensor nodes in the considered scenarios is performed using new MOGA-based. For the comparative analysis, a popular clustering method, namely evolutionary routing protocol (ERP) [23], is simulated on the same network setting. To compare the performance of each method, various performance matrices have been considered. In this chapter, we analyze the stability period, average residual energy of overall alive network, and overall lifetime of the network for both of the considered methods.

In the above defined simulation environment, the overall lifetime is measured in term of rounds. One round in the network is defined as the execution of one complete

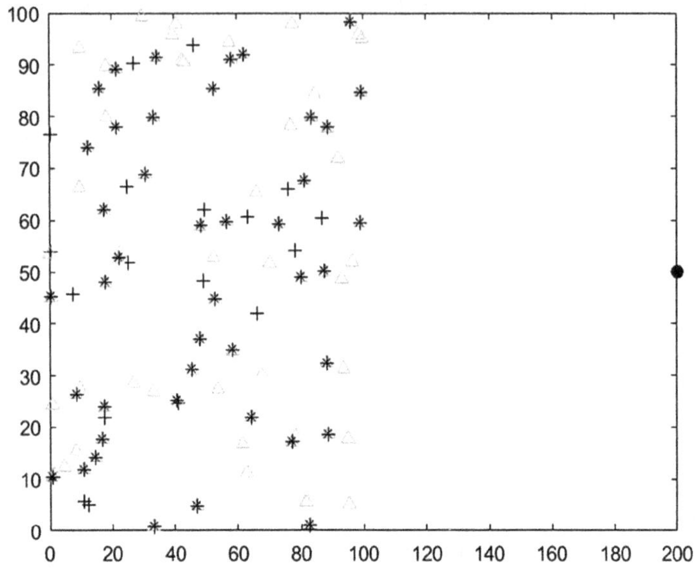

FIGURE 11.1 Sensor nodes after some rounds of operations.

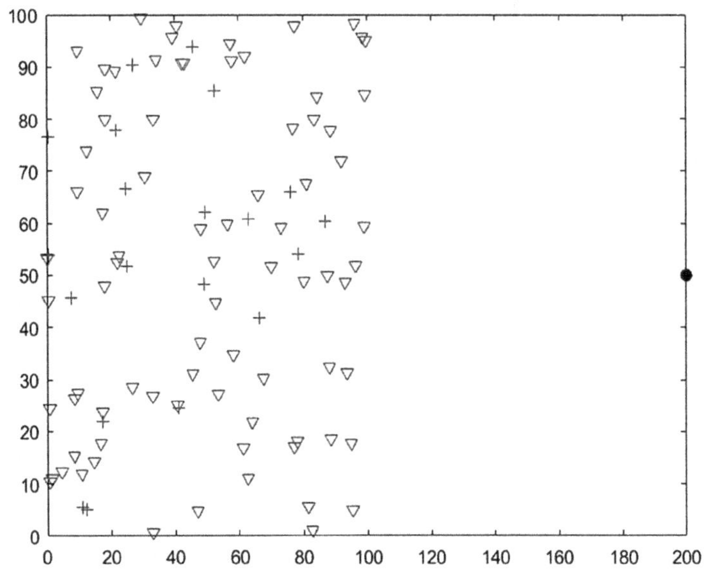

FIGURE 11.2 Dead sensor nodes after some rounds of operations.

cycle of setup and steady-state phases. In every round, the sink received some data from the network. The overall lifetime of the network can be defined as the number of rounds the network will take to exhaust all of its energy completely. If the network contains a greater number of advanced nodes, then the lifetime would be increased as

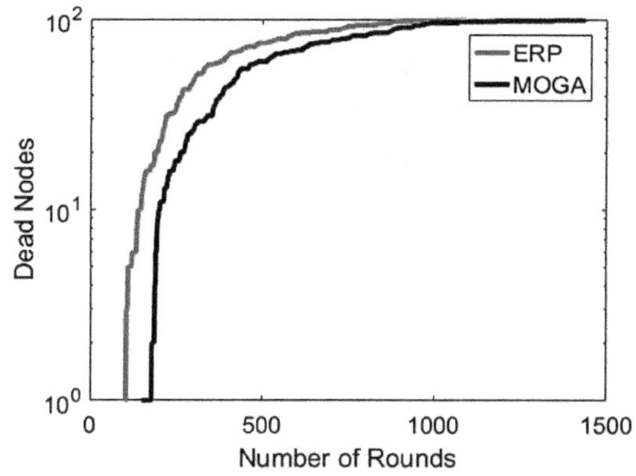

FIGURE 11.3 Scenario 1: Overall network lifetime with 10% advanced node.

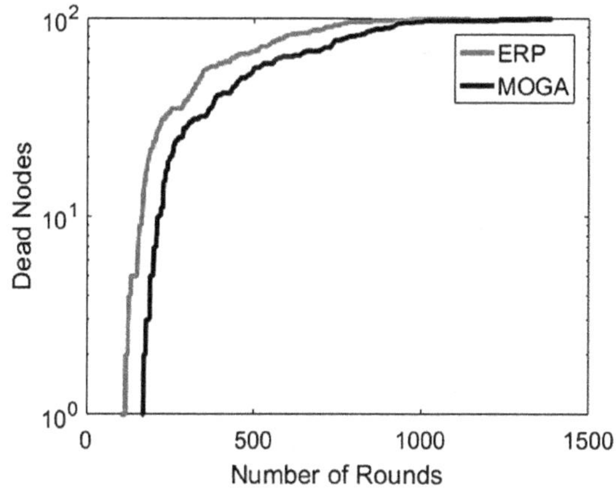

FIGURE 11.4 Scenario 2: Overall network lifetime with 20% advanced node.

compared to the network with a smaller number of advanced nodes. However, having a large number of advanced nodes will increase the cost of the overall network design. Therefore, in this chapter only two heterogeneity scenarios are presented. Figures 11.3 and 11.4 depict the overall lifetime comparison of ERP- and MOGA-based clustering methods for 10% and 20% advanced nodes in the network. In each of the scenarios, MOGA shows better performance than ERP as it took large number of rounds in depleting all of its energy.

Further, the same performance of both methods is presented quantitatively in Table 11.1 and Table 11.2. In ERP, the nodes are frequently as compared to MOGA from the start of the network operations. We have simulated the network for 100 sensor nodes. For scenario 1, the first 10 nodes had died by the 203 round in the case of MOGA while in the case of ERP, it will take 145 rounds. Moreover, the overall lifetime of the MOGA in this scenario is 1309 rounds while for ERP, it is 1093 rounds. Hence, MOGA outperforms ERP in scenario 1. For scenario 2, also

TABLE 11.1
Scenario 1: Dead Nodes Vs Number of Rounds

% Dead Node	ERP	MOGA
10	145	203
20	197	273
30	220	337
40	265	382
50	305	426
60	379	512
70	444	415
80	571	752
90	739	897
100	1093	1309

TABLE 11.2
Scenario 2: Dead Nodes Vs Number of Rounds

% Dead Node	ERP	MOGA
10	166	221
20	187	255
30	223	315
40	305	387
50	337	475
60	428	558
70	514	706
80	593	781
90	727	913
100	1145	1391

the lifetime of MOGA is 1391 rounds which is better than ERP, which has a lifetime of 1145 rounds.

Another important parameter for the performance comparison is the overall residual energy of the network. This can be calculated as the summation of remaining energy of all nodes in the current round of the network. Therefore, a graph is present for overall residual energy versus the number of rounds to analyze and compare the trends of both methods. Figures 11.5 and 11.6 presented the energy depletion behavior for both scenarios, respectively. From the visualization of both the curves, the performance of MOGA seems better than ERP. However, it looks similar for both the scenarios. Therefore, the same is further compared by quantitative analysis. Table 11.3 and 11.4 present result for scenario 1 and scenario 2 respectively till 1309 rounds only. The overall start energy of the network is 54.81 Joules and 59.79 Joules for both scenarios, respectively. The overall remaining energy is captured for different intervals of rounds and tabulated in these tables. It can easily be visualized that

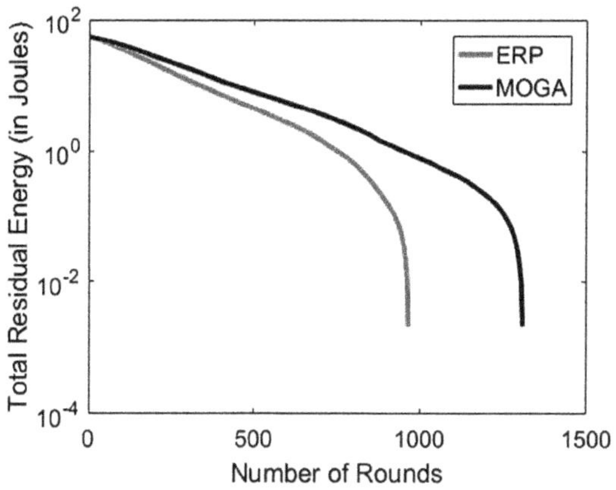

FIGURE 11.5 Scenario 1: Overall remaining energy for MOGA and ERP methods.

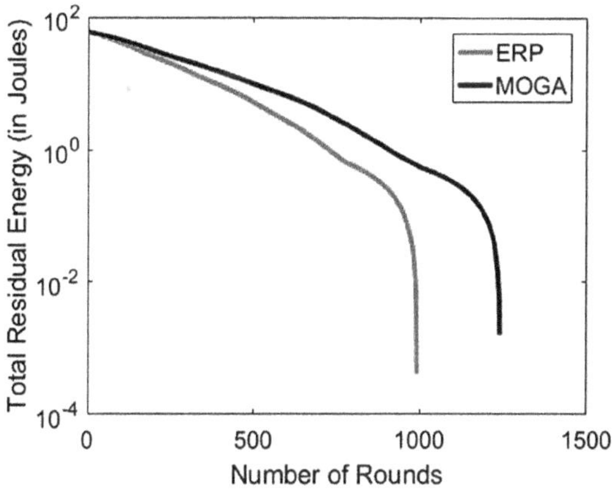

FIGURE 11.6 Scenario 2: Overall remaining energy for MOGA and ERP methods.

the MOGA based clustering method consumed less energy in each round and therefore the overall remaining energy of this method is higher than ERP method.

11.5 CONCLUSION

Clustering is the major application to address the crucial problem of extending the life of IWSNs, which are limited by low-capacity batteries. This chapter introduced a multi-objective approach using MOGA to do clustering of sensor nodes deployed randomly in any industrial scenario. To measure the quality of clustering, two objectives, namely compactness and separation, are considered as the fitness functions. The simulation results show that the proposed protocol outperformed in terms of several parameters: stability period, network life time, and residual energy. In future work, different hybrid approaches to clustering will be explored. The result of the proposed work will be validated with more sensor nodes and more sensor networks. This protocol was designed for stationary nodes.

TABLE 11.3
Scenario 1: Overall Remaining Energy Vs Number of Rounds

% Rounds	ERP	MOGA
0	54.81	54.85
10	30.69	37.09
20	14.9	21.6
30	7.5	12.1
40	3.97	7.25
50	1.99	4.42
60	0.74	2.47
70	0.12	1.19
80	0	0.61
90	0	0.25
100	0	0

TABLE 11.4
Scenario 2: Overall Remaining Energy Vs Number of Rounds

% Rounds	ERP	MOGA
0	59.79	59.87
10	34.23	39.82
20	17.53	23.47
30	8.53	13.97
40	3.57	7.79
50	1.31	4.03
60	0.46	1.66
70	0.04	0.66
80	0	0.29
90	0	0.02
100	0	0

REFERENCES

[1] M. Haenggi, A. I. Reuther, J. I. Goodman, D. R. Martinez, L. B. Ruiz, J. Nogueira, A. Loureiro, A. Boulis, B. Warneke, A. E. Kamal et al., *Handbook of Sensor Networks: Compact Wireless and Wired Sensing Systems*, CRC Press, 2005.

[2] R. Pal, S. Yadav, and R. Karnwal, "Eewc: Energy-efficient weighted clustering method based on genetic algorithm for hwsns," *Complex & Intelligent Systems*, vol. 6, no. 2, pp. 1–10, 2020.

[3] A. Cenedese, M. Luvisotto, and G. Michieletto, "Distributed clustering strategies in industrial wireless sensor networks," *IEEE Transactions on Industrial Informatics*, vol. 13, no. 1, pp. 228–237, 2016.

[4] K. Mehta and R. Pal, "Energy efficient routing protocols for wireless sensor networks: A survey," *International Journal of Computer Applications*, vol. 165, no. 3, pp. 41–46, May 2017.

[5] C.-Y. Chong and S. P. Kumar, "Sensor networks: Evolution, opportunities, and challenges," *Proceedings of the IEEE*, vol. 91, pp. 1247–1256, 2003.

[6] S. Bhushan, R. Pal, and S. G. Antoshchuk, "*Energy efficient clustering protocol for heterogeneous wireless sensor network: A hybrid approach using ga and k-means*," In *2018 IEEE Second International Conference on Data Stream Mining & Processing (DSMP)*, Lviv, Ukraine. IEEE, 2018, pp. 381–385.

[7] R. Pal, R. Sindhu, and A. K. Sharma, "*Sep-e (rch): Enhanced stable election protocol based on redundant cluster head selection for hwsns*," In *International Conference on Heterogeneous Networking for Quality, Reliability, Security and Robustness*. Greater Noida, India, Springer, 2013, pp. 104–114.

[8] R. Pal and A. K. Sharma, "*Fsep-e: Enhanced stable election protocol based on fuzzy logic for cluster head selection in Wsns*," In *Proc. of International Conference on Contemporary Computing*, Noida, India, IEEE, 2013.

[9] R. Pal, H. M. A. Pandey, and M. Saraswat, "*BEECP: Biogeography optimization-based energy efficient clustering protocol for HWSNs*," In *2016 Ninth International Conference on Contemporary Computing (IC3)*, Noida, India, IEEE, Aug 2016.

[10] K. Mehta and R. Pal, "*Biogeography based optimization protocol for energy efficient evolutionary algorithm: (BBO: EEEA)*," In *2017 International Conference on Computing and Communication Technologies for Smart Nation (IC3TSN)*, Faridabad, India, IEEE, Oct 2017.

[11] W. B. Heinzelman, A. P. Chandrakasan, and H. Balakrishnan, "An application-specific protocol architecture for wireless microsensor networks," *IEEE Transactions on Wireless Communications*, vol. 1, no. 4, pp. 660–670, 2002.

[12] G. Smaragdakis, I. Matta, and A. Bestavros, "*Sep: A stable election protocol for clustered heterogeneous wireless sensor networks*," In *Proceedings of International Workshop On Sensor and Actor Network Protocols and Applications*, Sweden, 2004.

[13] R. Pal and A. K. Sharma, "Msep-e: Enhanced stable election protocol with multihop communication," *Global Journal of Computer Science and Technology*, vol. 15, no. 8, 2014.

[14] V. Nehra, R. Pal, and A. K. Sharma, "Fuzzy-based leader selection for topology controlled pegasis protocol for lifetime enhancement in wireless sensor network," *International Journal of Computers & Technology*, vol. 4, no. 3, pp. 755–764, 2013.

[15] O. Younis and S. Fahmy, "Heed: A hybrid, energy-efficient, distributed clustering approach for ad hoc sensor networks," *IEEE Transactions on Mobile Computing*, vol. 3, pp. 366–379, 2004.

[16] R. Pal and M. Saraswat, "*Data clustering using enhanced biogeography-based optimization*," In *2017 Tenth International Conference on Contemporary Computing (IC3)*, Noida, India, IEEE, Aug 2017.

[17] R. Gupta and R. Pal, "*Biogeography-based optimization with léVY-flight exploration for combinatorial optimization*," In *2018 8th International Conference on Cloud Computing, Data Science & Engineering (Confluence)*, Noida, India, IEEE, Jan 2018.

[18] R. Pal and M. Saraswat, "Improved biogeography-based optimization," *International Journal of Advanced Intelligence Paradigms*, 2017.

[19] R. Pal, H. Mittal, and M. Saraswat, "*Optimal fuzzy clustering by improved biogeography-based optimization for leukocytes segmentation*," In *2019 Fifth International Conference on Image Information Processing (ICIIP)*, Waknaghat, India, IEEE, 2019, pp. 74–79.

[20] R. Pal and M. Saraswat, "Histopathological image classification using enhanced bag-of-feature with spiral biogeography-based optimization," *Applied Intelligence*, vol. 49, no. 09, pp. 1–19, 2019.

[21] E. Zitzler and L. Thiele, "Multiobjective evolutionary algorithms: A comparative case study and the strength pareto approach," *IEEE transactions on Evolutionary Computation*, vol. 3, no. 4, pp. 257–271, 1999.

[22] A. W. Matin and S. Hussain, "Intelligent hierarchical cluster-based routing," *Life*, vol. 7, p. 8, 2006.

[23] A. A. Bara and E. A. Khalil, "A new evolutionary based routing protocol for clustered heterogeneous wireless sensor networks," *Applied Soft Computing*, vol. 12, pp. 1950–1957, 2012.

[24] Q. Zhang and H. Li, "MOEA/d: A multiobjective evolutionary algorithm based on decomposition," *IEEE Transactions on Evolutionary Computation*, vol. 11, no. 6, pp. 712–731, Dec 2007.

[25] Pal, R., Saraswat, M., and Mittal, H., "Improved bag-of-features using grey relational analysis for classification of histology images," *Complex & Intelligent Systems*, vol. 7, no. 3, pp. 1429–1443, 2021.

[26] Mittal, H., Pandey, A. C., Pal, R., and Tripathi, A., "A new clustering method for the diagnosis of CoVID19 using medical images". *Applied Intelligence*, vol. 51, no. 5, pp. 2988–3011, 2021.

[27] Mittal, H., Tripathi, A., Pandey, A. C., and Pal, R., "Gravitational search algorithm: A comprehensive analysis of recent variants", *Multimedia Tools and Applications*, vol. 80, no. 5, pp. 1–28, 2020.

[28] Saraswat, M., Pal, R., Singh, R., Mittal, H., Pandey, A., and Bansal, J. C., "*An optimal feature selection approach using IBBO for histopathological image classification*", In *Congress on Intelligent Systems* (pp. 31–40). Springer, Singapore, 2020.

[29] Pal, R., Mittal, H., Pandey, A., and Saraswat, M., "An Efficient Bag-of-Features for Diseased Plant Identification", In Uddin M.S., Bansal J.C. (eds), *Computer Vision and Machine Learning in Agriculture*, pp. 159–172, 2021.

[30] Mittal, H., Saraswat, M., and Pal, R, "*Histopathological image classification by optimized neural network using igsa*", In *International Conference on Distributed Computing and Internet Technology* (pp. 429–436). Springer, Cham, 2020.

12 Monitoring Social Distancing for Industries and in Public Areas Using Machine Learning

Megha Agarwal, Vinti Gupta, and Abhinav Goel

CONTENTS

12.1 INTRODUCTION: BACKGROUND AND DRIVING FORCES

The time of these pandemics prompted a portion of the prudent steps the nation should take overall. One of the significant insurances that ought to be taken in these circumstances is Social Distancing and Quarantine. Specialists are doled out to save a check for public social occasions to not happen. Be that as it may, manual discovery by visiting each public spot is certainly not an incredible arrangement as it takes a lot of time. Currently, India is the worst affected country by COVID-19 and the death ratio is increasing day by day. To date, we have more than four vaccines available, but social distancing remains the best way to overcome and fight against this disease. In order to ensure social distancing protocols in overcrowded places and workplaces, this tool, which can monitor whether or not people are maintaining safe distancing protocols from one another by analyzing real-time video streams with the help of a camera. To keep track of people at various workplaces, factories, and shops we can use this tool in combination with their security camera systems and can monitor whether or not people are keeping a secure distance from one another. This chapter proposes a Machine Learning and

DOI: 10.1201/9781003089636-12

Python-based framework for monitoring social distancing using the surveillance video with the help of a security camera. In this proposed framework, we are utilizing the YOLOv3, an object detection model for segregation of humans being from the background and OpenCV for tracking the humans by using the bounded boxes and assigning IDs to them.

The most effective current method to overcome the spread of this deadly Covid-19 virus is to ensure social distancing. Social distance is guaranteed as the best spread prevention in the current situation, and all influenced nations are secured to execute social distancing [1]. The social distancing is not a very new concept for our society, because a lot of societies have been aware of it and have, for many generations, known the value of keeping a safe distance from a person who is suffering from any infectious disease. Social distancing is proved as the best method to stop the spread of this COVID-19 virus for the current situation. In 2020, we have seen that all affected countries were imposing a lockdown for the implementation of social distancing and that this proved as a boon for them to emerging out from this virus. Currently, India is suffering badly from COVID-19 as the second wave of COVID-19 is affecting the youth of the nation both physically and economically and that is the main reason for concern. The main objective for this research is to support and reduce the transmission rate, and the number of cases spreading over a longer time to relieve the pressure on our healthcare system, along with the minimum loss of economic endeavors. We have seen that maintaining a distance of about one metre from another person will help us to reduce transmission rate of most flu virus strains, including COVID-19.

In practice, this proves that avoiding close contact with one another will help us in slowing the rate of spread of infectious diseases like COVID-19. Social distancing is the only one of the non-pharmaceutical infection control methods that will help us to stop or slow down the spread of this highly contagious disease. When a healthy or uninfected person comes into contact with the respiratory droplets coming from the coughs or sneezes of an infected person, they can catch the infection and this virus can have a direct impact on the respiratory system. Currently, In India the situation is very critical and the demand for oxygen is increasing daily because the infected person is unable to inhale properly. So, to battle with this disease, this research will be beneficial. By using this detection tool, we can ensure the social distancing among people. This research uses the YOLO v3 Object Detection algorithm to detect the person and OpenCV is used for the tracking of the person and the assignment of a particular ID. The distance between the two bodies was then calculated using the traditional Euclidean Distance Formula given below [2].

$$d = \sqrt{\left[(x_2 - x_1)^2 + (y_2 - y_1)^2\right]}.$$

Then, by using the Euclidean distance formula we can calculate the distance between two people and figure out whether or not they are following the social distancing protocol or not (Figure 12.1).

FIGURE 12.1 Social distancing among peoples.

12.2 LITERATURE REVIEW

A substantial amount of research work was carried out for social distancing using different techniques and approaches.

Bielecki et al. [1] conducted a study of about 508 male soldiers, having an average age of 21 years. Initially, they divided the number of soldiers into two separate groups. Before the implementation of social distancing 354 soldiers were affected and 30% people make sick due to COVID-19. While no other soldier in a group of 154 was affected because of infections, after introducing social distancing among them. So, this proves the effect of implementing social distancing to overcome the spread of the COVID-19 virus.

N. Singh Punn et al. [2] proposes YOLO v3 object location model to isolate people from the background and the Deepsort method to deal with tracking the distinguished individuals with the assistance of bounding boxes and IDs. The results of the YOLO v3 model are further compared with other models, for example faster region based CNN (convolution neural network) and single shot locator (SSD) as far as of mean average precision (mAP), frames per second (FPS) and loss values defined by object classification and localization.

Ghorai et al. [3], proposed an advanced arrangement utilizing a Deep Learning strategy which would alarm them when the infringement of the social distancing is recognized; that is, number of individuals more than (as far as possible on the quantity of individuals permitted to be in a spot, set by the public authority) or distance between two individuals is not exactly the limit distance. A video stream will be caught from the CCTV camera. With the assistance of the PoseNet model we are distinguishing the people and monitoring the quantity of people present in the given live video transfer. If the quantity of people crosses the least edge limit (set by the authorities) or on the other hand if the Euclidean distance between any two poses detected in the frame recognized less than 3ft we alert the experts in-charged.

Yang et al. [4] have proposed an Artificial Intelligence (AI) monocular real-time camera-based system for monitoring social distancing. This system uses a social density for avoiding overcrowding in public areas by modulating now to the region of interest. In this method, verification was done using three different pedestrian crowd datasets. But the train station dataset was having some missing detections in it, because in some regions the density of peoples is very high and occlusion happens there. However, after doing some research on it, they decided that the maximum peoples who were captured there, for them the idea for social distance is valid.

Zheng et al. [5] note that distinguishing human bodies in profoundly packed scenes is a difficult issue. Two primary reasons bring about such an issue: 1) frail obvious signals of intensely impeded examples can scarcely give adequate data to precise identification; 2) vigorously blocked occasions are simpler to be stifled by Non-Greatest Concealment (NMS). In order to address these two issues, we present a variation of two-stage locators called S-RCNN. PS-RCNN first recognizes somewhat/none blocked items by a R-CNN [1] module (alluded as P-RCNN), and then, at that point, stifles the identified examples by human-formed veils so that the provisions of vigorously impeded examples can stand out. From that point onwards, PS-RCNN uses another R-CNN module had practical experience in intensely impeded human recognition (alluded as S-RCNN) to distinguish the rest missed items by P-RCNN. End-product are the gathering of the yields from these two RCNN.

Yadav et al. [6] developed a system which used the raspberry pi version 4 and integrated it with a camera for tracking social distancing in real time to overcome the spread of the virus. This model was trained on the custom dataset, which was installed in the raspberry pi 4 and there was a camera fitted into the model. The use of that camera was to feed it with real-time videos of the public places to the model in the raspberry pi 4, it will automatically and continuously monitor public places so that it can detect whether or not people in public places are maintaining a safe social distance and it also establishes whether or not those people are wearing masks. Their method occurs in two phases: First, if a person is not wearing a mask then his photo was taken and it will be sent to a control center at the police headquarters; and second, when the peoples violate the social distancing rules because it was detected continuously in threshold time then an alarm will ring out that orders people to maintain a social distance. This critical alert was sent to the control center of the police headquarters for taking action against peoples who was not following social distancing rules.

Chen et al. [7] focused on faster RCNN that detect the object. Author find the gap between NMS and POW.

Brunetti et al. [8] find the difference between 2D and 3D vision systems as well as indoor and outdoor systems. Author focused on the importance of testing pedestrian detection systems that apply on different datasets

Sener et al. [9] proposed another method. In this study, the motion of people who are communicating with each other was extracted from each region. After that, the visual descriptors were created for those two persons who were communicating with each other. Because the relative positions of those peoples are likely to complement these visual descriptors then we decided to use the embedded spatial multiple instances, that implicitly integrates the distance between those communicating people into the learning process of multiple instances. Experiments carried on two benchmark datasets and their endings validate that the use of two-person visual descriptors, along with more than one instance of spatial learning, provides them with an efficient way to infer the form of interaction. They achieved an accuracy of 83.3%.

12.3 PROPOSED METHODOLOGY

The proposed methodology focuses on ensuring social distancing among peoples, especially at overcrowded places. This system identifies the person with the help of camera attached to it for real-time monitoring and it can also detect social distancing on a recorded video or on the captured photo with the help of Machine Learning and Object Detection (Yolo V3). The proposed system will use the following algorithm for the detection of social distancing (Figure 12.2).

1. Object Detection Using YOLO
2. Object Tracking Using OpenCV
3. Distance Measurement.

12.3.1 Object Detection Using YOLO

The full form of YOLO is You Only Look Once. This algorithm is used to detect the objects in real time, and it is the most efficient and effective algorithm which is used for object detection. It also has many innovative ideas, which are emerging out from the computer vision research community. It also uses a fully different approach. For the detection of the objects in real time, this algorithm used a very clever Convolutional Neural Network (CNN). On a full image, YOLO will apply a single

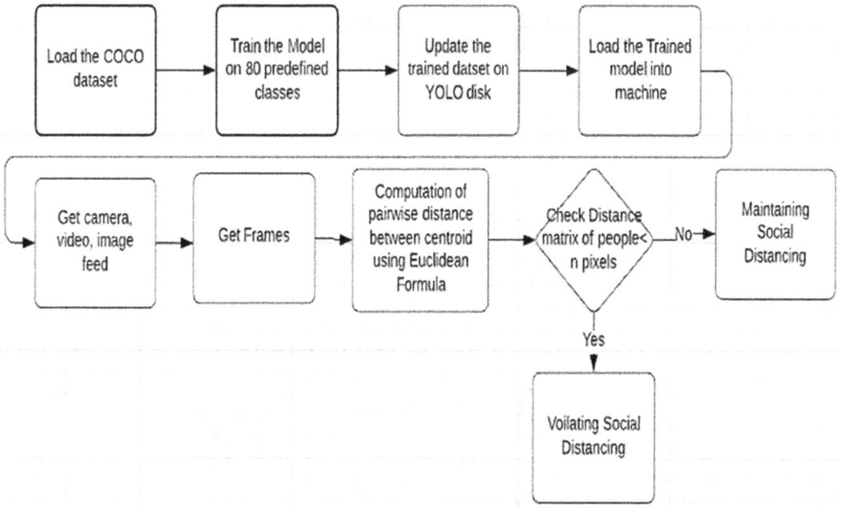

FIGURE 12.2 Architectural design of model.

FIGURE 12.3 Human Detection using YOLO.

neural network; after that, it will divide the image into various regions. Subsequently, it predicts about bounded boxes and the probabilities of every region. Those bounded boxes are weighted with the help of these predicted probabilities. In this research we are using a pre-trained YOLO v3 coco dataset for the model training. It was trained on more than 80 classes such as cat, dog, person, cars etc.; for our research, however, we are using only the person class for the detection of human being in the video or photo. The following steps are used for the training model (Figure 12.3).

1. First, we will load the COCO class labels with the help of the coco.names file, using this file our YOLO model will be trained for detecting images.
2. Then we will configure the path of the YOLO weights and its model configuration using weights and configuration file of the COCO dataset.
3. Finally, we will load our object detector, i.e. YOLO, which was trained on the COCO dataset (80 classes).

12.4 OBJECT TRACKING USING OPENCV

OpenCV is an open-source library for image processing, computer vision and Machine Learning. OpenCV plays an important role in real-time operation and for today's systems it is very important. With the help of OpenCV, we can process images and videos in order to identify the faces, objects, or even handwriting of a human. After the detection of the human being we have to keep track of the movement done by the human being in the video. To achieve this goal, OpenCV is the best tool. It will basically assign a particular ID to the human being tracked in the frame. And with the help of these IDs assigned to a particular we can keep track of the movement done by that particular person. OpenCV has an advanced feature by which it needs the phase of object detection only once, i.e. when that object is first detected. OpenCV is a very fast algorithm in terms of object tracking. When any object moves or disappears outside the boundaries in the given video frame, OpenCV will be able to handle it easily. Also, it will be capable of picking up those objects that has been lost from the frames in between (Figures 12.4–12.7).

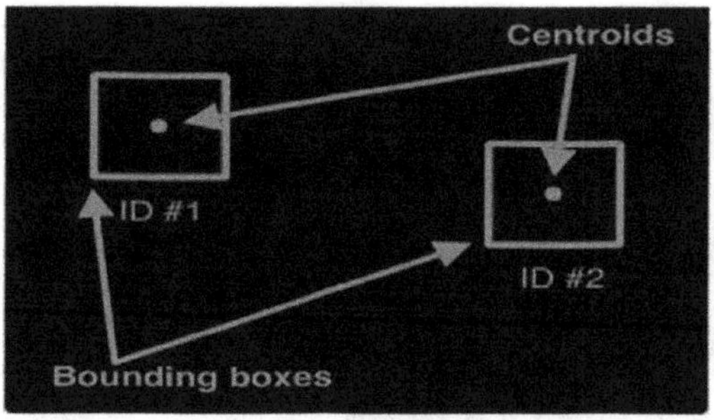

FIGURE 12.4 Use of Centroid Tracking to create an object tracking algorithm.

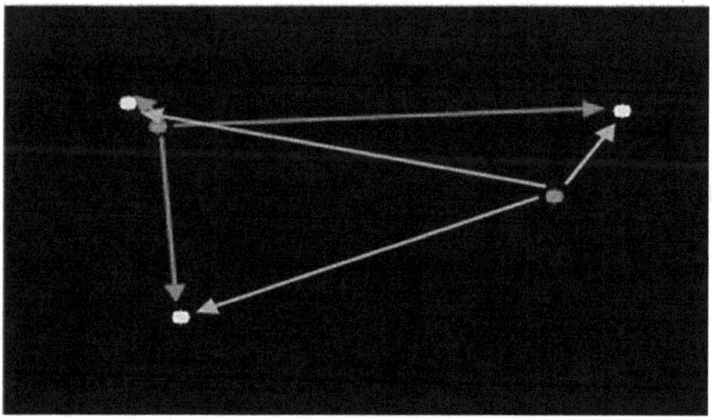

FIGURE 12.5 Tracking of objects changing their position using OpenCV and Python. Here, we will calculate the Euclidean distances between each pair of new centroids (*green*) and original centroids (*red*).

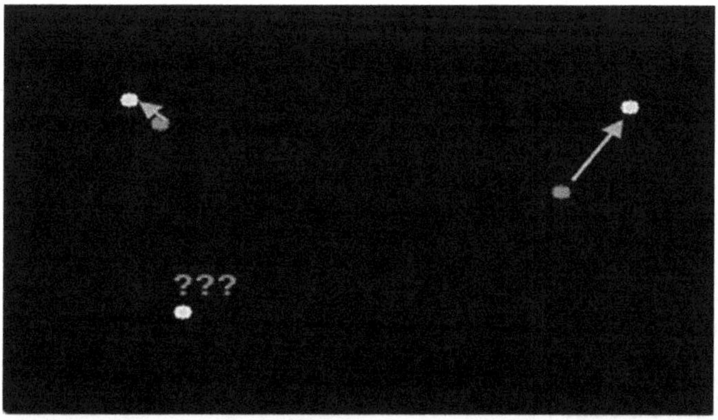

FIGURE 12.6 In this figure, our centroid object tracking method is having objects associated with minimized distance of objects. So, what ID we have to assign it?

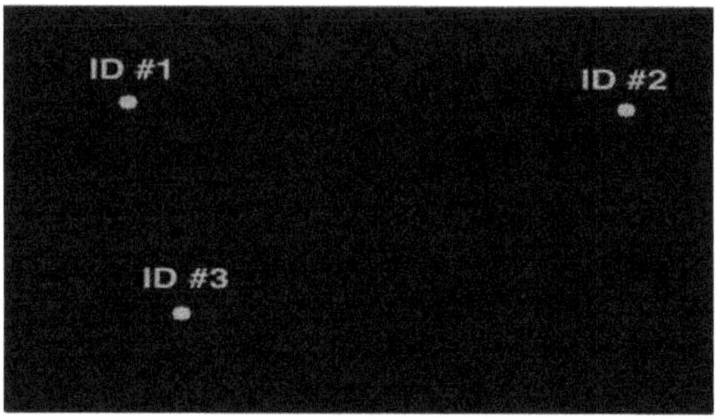

FIGURE 12.7 Using Object tracking with OpenCV and Python example, here a new object is found that is not matching with previous object, so a new ID is given to that object as ID #3.

12.4.1 Distance Measurement

This is the final stage of the research working. It is the main phase of the research because here we will check whether or not the social distancing protocol is being followed by the peoples. Here, with the help of the Euclidean distance formula we will calculate the distance between the peoples present with the help of those IDs provided by the OpenCV. We will then calculate the distance between all pairs of centroids in the plane. We will check whether or not the distance between two centroids of the nearest boxes is less than the configured number of pixels. If the distance between the pair of centroids is less than the configured number of pixels then an alert will be generated and the color of boxes in which these centroids are present will be changed to red. If any index pair that exists within the violation/abnormal sets, then it will update that color to yellow. If the box color is yellow, then it means that the violation is abnormal and if that color is red then this will indicate a critical violation and an alert will be generated.

12.5 OUTCOMES

The outcomes of this research will be as follows (Figure 12.8):

1. It will count the number of peoples present in that video frame by using Objection Detection.
2. It will calculate the distance between each, showing how many people are following the social distancing protocols.
3. It will also display the total number of serious violations and also the number of abnormal violations.
4. It will also display the minimum safe distance to be kept between one another.

12.6 CONCLUSION

This research will help us to overcome the spread of this deadly COVID-19 virus. Currently, the situation is very critical and we all have to keep social distancing in

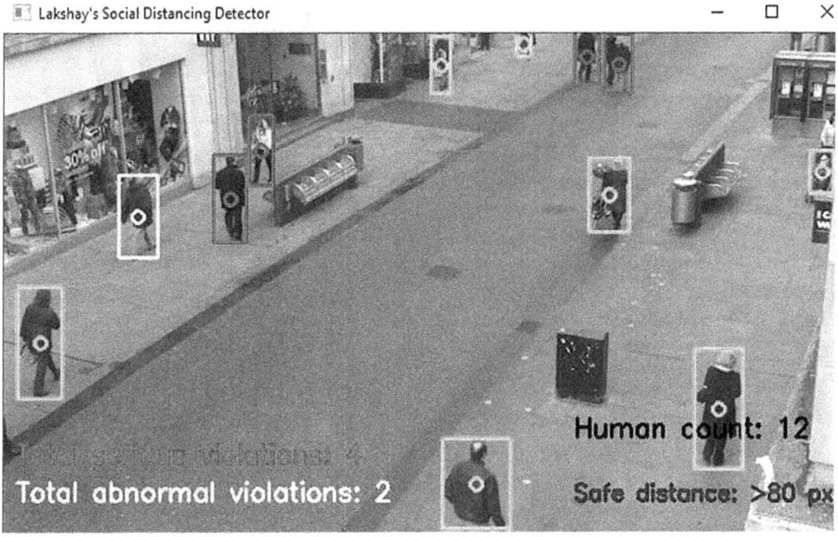

FIGURE 12.8 Final outcome of the research.

these tough times. It will help us to ensure social distancing in overcrowded places such as cinema halls and shopping malls by real-time monitoring with help of surveillance cameras. Additionally, we can figure out social distancing on a recorded video and check whether peoples are following social distancing protocols as per government guidelines with the help of bounding boxes for object detection and if they are not following proper protocols then strict action can be taken against them. So, it will help the government to find out these anti-social elements who are putting mankind in danger. This social distancing detector will contribute to public healthcare and help mankind.

In future, we can send alert message to authorities if rules are violated. So the authorities can take action immediately. It is easy to detect that the social distancing whether or not the protocol is followed.

We can detect temperature, cough and sneezing as they are the major symptoms for Covid-19 using various Deep Learning Techniques so that our system work more efficiently in detecting Covid-19 cases.

REFERENCES

1. Bielecki, Michel, Roland Züst, Denise Siegrist, Daniele Meyerhofer, Giovanni Andrea Gerardo Crameri, Zeno Stanga, Andreas Stettbacher, Thomas Werner Buehrer, and Jeremy Werner Deuel. "Social distancing alters the clinical course of COVID-19 in young adults: A comparative cohort study." *Clinical Infectious Diseases* 72, no. 4 (2021): 598–603.
2. Punn, Narinder Singh, Sanjay Kumar Sonbhadra, Sonali Agarwal, and Gaurav Rai. "Monitoring COVID-19 social distancing with person detection and tracking via fine-tuned YOLO v3 and Deepsort techniques." arXiv preprint arXiv:2005.01385 (2020).
3. Ghorai, Arnab, Sarah Gawde, and Dhananjay Kalbande. "*Digital solution for enforcing social distancing.*" In *Proceedings of the International Conference on Innovative Computing & Communications (ICICC)*. 2020.

4. Yang, D., E. Yurtsever, V. Renganathan, K. A. Redmill, and Ü. Özgüner. "A vision-based social distance and critical density detection system for COVID-19." (2020).
5. Ge, Zheng, Zequn Jie, Xin Huang, Rong Xu, and Osamu Yoshie. "*PS-RCNN: Detecting secondary human instances in a crowd via primary object suppression.*" In *2020 IEEE International Conference on Multimedia and Expo (ICME)*, pp. 1–6. IEEE, 2020.
6. Yadav, Shashi. "Deep learning based safe social distancing and face mask detection in public areas for Covid-19 safety guidelines adherence." *International Journal for Research in Applied Science and Engineering Technology* 8, no. 7 (2020): 1368–1375.
7. Chen, X. and A. Gupta. "An implementation of faster RCNN with study for region sampling." arXiv preprint arXiv:1702.02138, (2017).
8. Brunetti, A., D. Buongiorno, G. F. Trotta, and V. Bevilacqua. "Computer vision and deep learning techniques for pedestrian detection and tracking: A survey." *Neurocomputing* 300 (2018): 17–33.
9. Sener, F. and N. Ikizler-Cinbis. "Two-person interaction recognition via spatial multiple instance embedding." *Journal of Visual Communication and Image Representation* 32 (2015): 63.

13 Profit-Maximization Inventory Model with Stock-Dependent Demand

Archana Sharma and A. K. Malik

CONTENTS

13.1 INTRODUCTION

Today's business organizers day invent new policies/ideas on a daily basis to increase customer demand, which will, in turn, increase company profits. Recently, many researchers have been focused on the development of an inventory model in a realistic environment. Over the past few years, a number of researchers have shown an interest in the economic production model (EPQ) too attached with day-to-day activities. Good-quality stocks are always attracting customers and this leads to higher demands for the products. In today's competitive markets, it is necessary to identify the pattern of these customers' demands. It has been seen that for attending enough customer service level in the current scenario, it is viable to have more stock. Therefore, in the present scenario, dealing with the stock-dependent demand model is most appropriate. Levin et al. (1972) noted that "large piles of consumer goods displayed in a supermarket will lead consumers to buy more. Yet too many goods piled up in everyone's way leave a negative impression on buyers and employees alike." In one study, Silver and Peterson (1985) discussed a model in which retail store for sales volume is directly proportional to the displayed items.

In order to secure high sales revenue, a firm should have high inventory level; however, this would increase the inventory holding cost due to larger inventory and cost of deterioration increase in product demand, along with higher purchasing costs. But it will achieve organization objectives to have higher profit. So, organizations try

DOI: 10.1201/9781003089636-13

to maximize the profit if the demand for these products is stock-dependent. There were many studies on the EPQ model related to a single item, whose demand is stock-dependent. Datta and Pal (1990) focused on demand taken as stock level for the proposed mode in that condition. Chung (2003) proposed an application of a stock-dependent demand in an inventory system. Later, an improved EPQ model considered with decaying products and stock-dependent demand is assumed by Teng and Chang (2005). Similarly, Alfares (2007) suggested a mathematical model system considered variable holding cost and demand dependence on the stock of the items.

Many models were developed for production-inventory model and determined optimum lot size which minimized the total cost of the entire supply chain management. Recently, the stock-dependent demand model has become increasingly admired among researchers, especially as it provides a tempting framework for the analysis of asset of the company. Some improved mathematical models, based on stock-dependent demand having various parameters, were developed by Singh and Malik (2011), Singh et al. (2011), Sarkar and Sarkar (2013) and Gupta et al. (2013) among others.

In today's era of improving business, the industries/companies demonstrates the product through malls/supermarkets which attract customer to purchase more and more products. A number of products in a large quantity and variety attract the customer and a result company experiences higher demand. Malik et al. (2016) demonstrated an improved inventory model without shortages and considered the time-varying and non-instantaneous products with a maximum lifetime. Kumar et al. (2016) analyse a mathematical model in which considered stock-dependent demand and variable holding cost a time function. Vashisth et al. (2016) proposed a multivariable demand inventory model assumed demand dependent on time and stock and assumed non-instant decaying products. Sharma et al. (2017) presented an inventory management system for deteriorating items in which ramp-type demand with inflation conditions. Shukla et al. (2017) proposed a model in two different categories in regard of holding costs and calculated total minimum inventory cost under ramp-type demand. Finally, Sharma et al. (2018) worked out for optimal policy for developed model with deteriorating items under the ramp-type demand.

In the past decade a number of researchers have been attracted to work with EPQ inventory models. Ruidas et al. (2020) emphasized a theory for a production model in which stock-price dependent demand assumed for decaying items where different cost parameters considered as interval numbers and find the profit function. Improved mathematical inventory models for non-instantaneous decaying products discussed by Malik et al. (2019) and Mathur et al. (2019) with variable constraints. Rout et al. (2020) extracted the scrapped items under the imperfect production process for the developed system. Saxena et al. (2020) demonstrated an improved model with selling price demand under two warehouses system. Rahaman et al. (2020) captured an EPQ model under uniform demand by using fractional calculus.

The proposed inventory model jointly discussed the stock-dependent demand and derived three cases arises due to stock-dependent demand rate. Then the model is solved numerically to get optimal replenishment cycle, which would maximize the total profit. The rest of the chapter is structured as follows. Section 13.2 covers the notation and assumptions. Section 13.3 introduces the mathematical model with the stock-dependent demand rate. Section 13.4 performs the solution procedure with some numerical examples. Section 13.5 develops the sensitivity analysis and finally Section 13.6 performs the conclusion and discusses the various areas for further research work.

13.2 NOTATIONS AND ASSUMPTIONS

The following notations and assumptions are used to propose this model with stock-dependent demand:

Notation:

- A = ordering cost is (per order).
- h = holding cost is (per unit time).
- C_p = purchase cost is (per unit time).
- C_s = selling price is (per unit time).
- P(t) = Production rate is (units/unit time) and it is demand dependent. P(t) = k R(I(t)) and k > 1.
- Q = maximum inventory level starting at t = 0.
- Q_0 = inventory level up to which demand rate is constant.
- I(t) = on-hand inventory-level for the proposed model at any time *t*.
- T = total cycle time.
- $Z_1(T)$ = the total profit function is (per unit time) for Case I.
- $Z_2(T)$ = the total profit function is (per unit time) for Case II.
- $Z_3(T)$ = the total profit function is (per unit time) for Case III.
- D = constant demand rate.

Assumptions:

1. The inventory model proposed for single product.
2. Replenishments are instantaneous at an infinite rate.
3. Demand rate R(I(t)) is deterministic and is taken as follows:

$$R(I(t)) = \begin{cases} \alpha[I(t)]^{\beta} & \text{if } I(t) > Q_0 \\ D & \text{if } 0 < I(t) < Q_0 \end{cases}$$

   where $\alpha > 0$, $0 < \beta < 1$, and α *and* β are called as scale and shape constraints.
4. The shortages are not allowed.
5. The delivery lead-time is zero.
6. The Production rate assumed as finite.

13.3 MATHEMATICAL FORMULATION

Here we developed a model for the various types of demand rate, which are constant and stock-dependent. We consider a production model in which consider the stock-dependent demand rate in diverse types of time intervals. There are three different situations may arise:

Case I: If $Q \le Q_0$, then demand is constant in this period and it is only the case of a classical EPQ model (Figure 13.1).

Case II: If $I(T) < Q_0 < Q$, then the demand initially is constant and after that it becomes stock-dependent after the inventory level extends to Q (Figure 13.2).

Case III: If $Q > I(T) > Q_0$, then the demand is assumed to be power form of on-hand inventory (Figure 13.3).

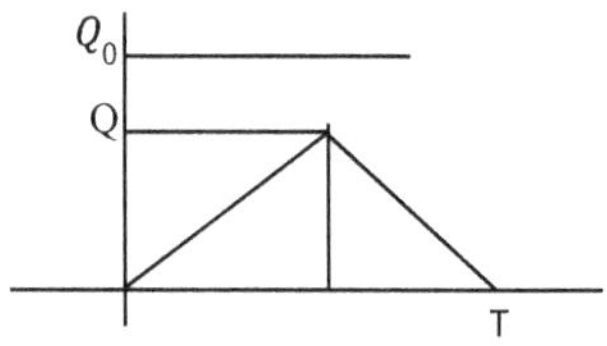

FIGURE 13.1 Pictorial representation of production system when $Q \le Q_0$.

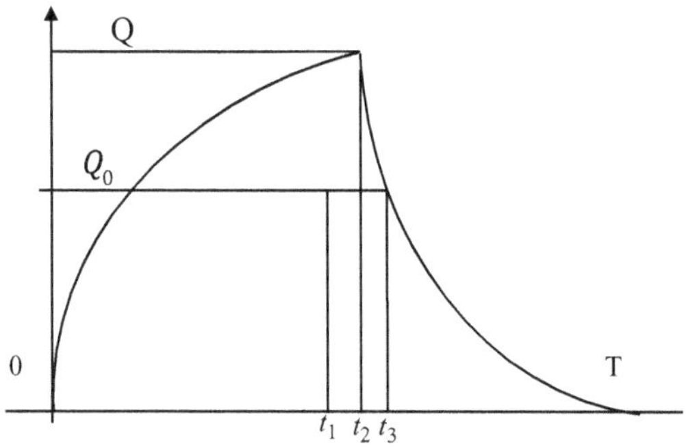

FIGURE 13.2 Pictorial representation of production system when $I(T) < Q_0 < Q$.

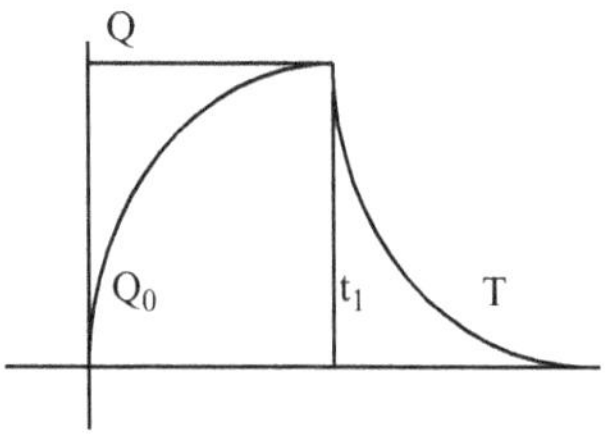

FIGURE 13.3 Pictorial representation of production system when $Q_0 < I(T)$.

Case I:

$Q \le Q_0$

At the time interval $[0, t_1]$, production occurs, and inventory increases by production, and demand. At time interval $[t_1, T]$, we fulfil the customer's demand so in this interval inventory affected by only demand.

As per above description, inventory level governed the succeeding equations:

$$I'(t) = P - D \quad 0 < t < t_1 \tag{13.1}$$

and

$$I'(t) = -D \quad t_1 < t < T \tag{13.2}$$

with the boundary conditions I(0) = 0 and I(T) = 0.

Solutions are

$$\mathrm{I}(\mathrm{t}) = (\mathrm{k}-1)\mathrm{D}\mathrm{t} \tag{13.3}$$

$$\mathrm{I}(\mathrm{t}) = \mathrm{D}(\mathrm{T}-\mathrm{t}) \tag{13.4}$$

The associated costs are:

$$\text{Holding cost}(\text{HC}) = \int_0^T hI(t)\,dt$$

$$= hD\left\{k\frac{t_1^2}{2} + \frac{T^2}{2} - Tt_1\right\}$$

Purchasing cost (PC) = C_pDT

Revenue cost (RC) = C_sDT

Ordering cost (OC) = A

Total profit function (per cycle) is represented by

$$\begin{aligned} Z_1(t_1,T) &= \frac{1}{T}\left[\text{RC} - \text{PC} - \text{HC} - \text{OC}\right] \\ &= \frac{1}{T}\left[(C_s - C_p)DT - hD\left\{k\frac{t_1^2}{2} + \frac{T^2}{2} - Tt_1\right\} - \text{A}\right] \end{aligned} \tag{13.5}$$

At $t = t_1$ from Equations (13.3) and (13.4), $t_1 = \dfrac{T}{k}$

Therefore,

$$Z_1(T) = \frac{1}{T}\left[(C_s - C_p)DT - hD\left\{\frac{T^2(k-1)}{2k}\right\} - \text{A}\right] \tag{13.6}$$

Our aim is to maximize the function $Z_1(T)$, using the essential conditions for $Z_1(T)$ to be maximize is $\frac{\partial Z_1(T)}{\partial T} = 0$ and $\frac{\partial^2 Z_1}{\partial T^2} < 0$ for all $T > 0$.

Case II:

$$I(T) < Q_0 < Q$$

The inventory level governed the succeeding equations:

$$I'(t) = P - D \quad 0 < t < t_1 \tag{13.7}$$

$$I'(t) = P - \alpha\left[I(t)\right]^{\beta} \quad t_1 < t < t_2 \tag{13.8}$$

$$I'(t) = -\alpha\left[I(t)\right]^{\beta} \quad t_2 < t < t_3 \tag{13.9}$$

$$I'(t) = -D \quad t_3 < t < T \tag{13.10}$$

Solutions of Equations (13.7), (13.8), (13.9) and (13.10) with boundary conditions $I(0) = 0$, $I(t_1) = Q_0$, $I(t_3) = Q_0$ and $I(T) = 0$

$$\mathrm{I}(\mathrm{t}) = (\mathrm{k}-1)\mathrm{D.t} \tag{13.11}$$

$$I(t) = \left[(k-1)(1-\beta)\alpha(t-t_1) + Q_0^{1-\beta}\right]^{\frac{1}{1-\beta}} \tag{13.12}$$

$$I(t) = \left[(1-\beta)\alpha(t_3-t) + Q_0^{1-\beta}\right]^{\frac{1}{1-\beta}} \tag{13.13}$$

$$\mathrm{I}(\mathrm{t}) = \mathrm{D}(\mathrm{T}-\mathrm{t}) \tag{13.14}$$

$$\text{Holding cost} = \int_0^T hI(t)\,dt = \int_0^{t_1} hI(t)\,dt + \int_{t_1}^{t_2} hI(t)\,dt + \int_{t_2}^{t_3} hI(t)\,dt + \int_{t_3}^{T} hI(t)\,dt$$

$$= (k-1)D\frac{t_1^2}{2} + \frac{h}{\alpha(k-1)(2-\beta)}\begin{bmatrix}\left\{(k-1)(1-\beta)\alpha(t_2-t_1) + Q_0^{1-\beta}\right\}^{\frac{2-\beta}{1-\beta}} \\ -Q_0^{2-\beta}\end{bmatrix}$$
$$-\frac{h}{\alpha(2-\beta)}\left[Q_0^{2-\beta} - \left\{(1-\beta)\alpha(t_3-t_2) + Q_0^{1-\beta}\right\}^{\frac{2-\beta}{1-\beta}}\right] + \frac{hD(T-t_3)^2}{2}$$

$$\text{Purchasing cost} = C_p\left[\int_0^{t_1} D\,dt + \int_{t_1}^{t_2} \alpha\left[I(t)\right]^{\beta} dt + \int_{t_2}^{t_3} \alpha\left[I(t)\right]^{\beta} dt + \int_{t_3}^{T} D\,dt\right]$$

$$= C_p\begin{bmatrix} Dt_1 + \frac{1}{k-1}\left(\left\{(k-1)(1-\beta)\alpha(t_2-t_1) + Q_0^{1-\beta}\right\}^{\frac{1}{1-\beta}} - Q_0\right) \\ -\left(Q_0 - \left\{\alpha(1-\beta)(t_3-t_2) + Q_0^{1-\beta}\right\}^{\frac{1}{1-\beta}}\right) + D(T-t_3)\end{bmatrix}$$

$$\text{Revenue cost} = C_s\left[\int_0^{t_1} D dt + \int_{t_1}^{t_2} \alpha\left[I(t)\right]^{\beta} dt + \int_{t_2}^{t_3} \alpha\left[I(t)\right]^{\beta} dt + \int_{t_3}^{T} D dt\right]$$

$$= C_s\left[\begin{array}{l} Dt_1 + \dfrac{1}{k-1}\left(\left\{(k-1)(1-\beta)\alpha(t_2-t_1)+Q_0^{1-\beta}\right\}^{\frac{1}{1-\beta}} - Q_0\right) \\ -\left(Q_0 - \left\{\alpha(1-\beta)(t_3-t_2)+Q_0^{1-\beta}\right\}^{\frac{1}{1-\beta}}\right) + D(T-t_3) \end{array}\right]$$

Ordering cost = A

Total profit function (per cycle) is represented by

$$Z_2(t_1,t_2,t_3T) = \frac{1}{T}\left[\text{RC} - \text{PC} - \text{HC} - \text{OC}\right]$$

$$= \frac{1}{T}(C_s - C_p)\left[\begin{array}{l} Dt_1 + \dfrac{1}{k-1}\left(Q_0 - \left\{(k-1)(1-\beta)\alpha(t_2-t_1)+Q_0^{1-\beta}\right\}^{\frac{1}{1-\beta}}\right) \\ -\left(Q_0 - \left\{\alpha(1-\beta)(t_3-t_2)+Q_0^{1-\beta}\right\}^{\frac{1}{1-\beta}}\right) + D(T-t_3) \end{array}\right]$$

$$-\frac{1}{T}(k-1)D\frac{t_1^2}{2} - \frac{h}{\alpha(k-1)(2-\beta)}\frac{1}{T}\left[\left\{(k-1)(1-\beta)\alpha(t_2-t_1)+Q_0^{1-\beta}\right\}^{\frac{2-\beta}{1-\beta}} - Q_0^{2-\beta}\right]$$

$$+\frac{h}{\alpha(2-\beta)}\frac{1}{T}\left[Q_0^{2-\beta} - \left\{(1-\beta)\alpha(t_3-t_2)+Q_0^{1-\beta}\right\}^{\frac{2-\beta}{1-\beta}}\right] - \frac{hD(T-t_3)^2}{2}\frac{1}{T} - A\frac{1}{T} \tag{13.15}$$

At t = t_1 from Equations (13.11) and (13.12), $t_1 = \dfrac{Q_0}{(k-1)D}$

At t = t_2 from Equations (13.12) and (13.13), $t_2 = \dfrac{T}{k}$

At t = t_3 from Equations (13.13) and (13.14), $t_3 = \dfrac{DT - Q_0}{D}$

Therefore,

$$Z_2(T) = \frac{1}{T}(C_s - C_p)\left[\begin{array}{l} Dt_1 + \dfrac{1}{k-1}\left(Q_0 - \left\{\begin{array}{l}(k-1)(1-\beta)\alpha\left(\dfrac{T}{k} - \dfrac{Q_0}{(k-1)D}\right) \\ +Q_0^{1-\beta}\end{array}\right\}^{\frac{1}{1-\beta}}\right) \\ -\left(Q_0 - \left\{\alpha(1-\beta)\left(\dfrac{DT-Q_0}{D} - \dfrac{T}{k}\right) + Q_0^{1-\beta}\right\}^{\frac{1}{1-\beta}}\right) \\ +D\left(T - \dfrac{DT-Q_0}{D}\right) \end{array}\right]$$

$$-\frac{1}{T}(k-1)D\left(\frac{Q_0}{(k-1)D}\right)^2\frac{1}{2}-\frac{h}{\alpha(k-1)(2-\beta)}\frac{1}{T}$$
$$\left[\left\{(k-1)(1-\beta)\alpha\left(\frac{T}{k}-\frac{Q_0}{(k-1)D}\right)+Q_0^{1-\beta}\right\}^{\frac{2-\beta}{1-\beta}}-Q_0^{2-\beta}\right]$$
$$+\frac{h}{\alpha(2-\beta)}\frac{1}{T}\left[Q_0^{2-\beta}-\left\{(1-\beta)\alpha\left(\frac{DT-Q_0}{D}-\frac{T}{k}\right)+Q_0^{1-\beta}\right\}^{\frac{2-\beta}{1-\beta}}\right]$$
$$-\frac{hD\left(T-\frac{DT-Q_0}{D}\right)^2}{2}\frac{1}{T}-A\frac{1}{T} \tag{13.16}$$

Our aim is to maximize the function $Z_2(T)$, using the essential conditions for $Z_2(T)$ to be maximize is $\frac{\partial Z_2(T)}{\partial T}=0$ and $\frac{\partial^2 Z_2}{\partial T^2}<0$ for all $T>0$.

Case III:

$$Q_0 < I(T)$$

From case-3, the equations leading the system as demonstrated:

$$I'(t)=P-\alpha\left[I(t)\right]^{\beta} \quad 0<t<t_1 \tag{13.17}$$
$$I'(t)=-\alpha\left[I(t)\right]^{\beta} \quad t_1<t<T \tag{13.18}$$

Solutions the Equations (13.17–13.18) using the boundary conditions,

$$\mathrm{I}(0)=\mathrm{Q}_0 \text{ and } \mathrm{I}(\mathrm{T})=0$$
$$I(t)=\left[(k-1)(1-\beta)\alpha t+Q_0^{1-\beta}\right]^{\frac{1}{1-\beta}} \tag{13.19}$$
$$I(t)=\left[(1-\beta)\alpha(T-t)\right]^{\frac{1}{1-\beta}} \tag{13.20}$$

$$\text{Holding cost}=\int_0^{t_0} hI(t)\,dt+\int_{t_0}^{T} hI(t)\,dt$$
$$=\int_0^{t_1} h\left[(k-1)(1-\beta)\alpha t+Q_0^{1-\beta}\right]^{\frac{1}{1-\beta}}dt+\int_{t_1}^{T} h\left[(1-\beta)\alpha(T-t)\right]^{\frac{1}{1-\beta}}dt$$
$$=\frac{h}{\alpha(k-1)(2-\beta)}\left[\left\{(k-1)(1-\beta)\alpha t_1+Q_0^{1-\beta}\right\}^{\frac{2-\beta}{1-\beta}}-Q_0^{2-\beta}\right]$$
$$+\frac{h}{\alpha(2-\beta)}\left[\left\{(1-\beta)\alpha(T-t_1)\right\}^{\frac{2-\beta}{1-\beta}}\right]$$

$$\text{Purchasing cost} = \frac{C_p}{k-1}\left[\left\{(k-1)(1-\beta)\alpha t_1 + Q_0^{1-\beta}\right\}^{\frac{1}{1-\beta}} - Q_0\right] + C_p\left[\alpha(1-\beta)(T-t_1)\right]^{\frac{1}{1-\beta}}$$

$$\text{Revenue cost} = \frac{C_s}{k-1}\left[\left\{(k-1)(1-\beta)\alpha t_1 + Q_0^{1-\beta}\right\}^{\frac{1}{1-\beta}} - Q_0\right] + C_s\left[\alpha(1-\beta)(T-t_1)\right]^{\frac{1}{1-\beta}}$$

Ordering cost = A

Total profit per cycle is given by

$$Z_3(t_1,T) = \frac{1}{T}\left[\mathrm{RC}-\mathrm{PC}-\mathrm{HC}-\mathrm{OC}\right]$$

$$= \frac{1}{T}\left[\begin{array}{l}(C_s-C_p)\left[\begin{array}{l}\left\{(k-1)(1-\beta)\alpha t_1 + Q_0^{1-\beta}\right\}^{\frac{1}{1-\beta}} - Q_0 \\ +(C_s-C_p)\left[\alpha(1-\beta)(T-t_1)\right]^{\frac{1}{1-\beta}} \\ -\dfrac{h}{\alpha(k-1)(2-\beta)}\left[\left\{(k-1)(1-\beta)\alpha t_1 + Q_0^{1-\beta}\right\}^{\frac{2-\beta}{1-\beta}}\right] \\ -Q_0^{2-\beta}\end{array}\right] \\ -\dfrac{h}{\alpha(2-\beta)}\left[\left\{(1-\beta)\alpha(T-t_1)\right\}^{\frac{2-\beta}{1-\beta}}\right] - A\end{array}\right] \tag{13.21}$$

At t = t_1 from Equations (13.19) and (13.20), $t_1 = \dfrac{T}{k} + \dfrac{Q_0^{1-\beta}}{\alpha(1-\beta)k}$

Therefore,

$$Z_3(T) = \frac{1}{T}\left[\begin{array}{l}(C_s-C_p)\left(\begin{array}{l}\left\{(k-1)(1-\beta)\alpha\left(\dfrac{T}{k}+\dfrac{Q_0^{1-\beta}}{\alpha(1-\beta)}k\right)+Q_0^{1-\beta}\right\}^{\frac{1}{1-\beta}} - Q_0 \\ +(C_s-C_p)\left[\alpha(1-\beta)\left(T-\left(\dfrac{T}{k}+\dfrac{Q_0^{1-\beta}}{\alpha(1-\beta)k}\right)\right)\right]^{\frac{1}{1-\beta}} \\ -\dfrac{h}{\alpha(k-1)(2-\beta)}\left[\left\{(k-1)(1-\beta)\beta\left(\dfrac{T}{k}+\dfrac{Q_0^{1-\beta}}{\alpha(1-\beta)k}\right)+Q_0^{1-\beta}\right\}^{\frac{2-\beta}{1-\beta}}\right] \\ -Q_0^{2-\beta}\end{array}\right) \\ -\dfrac{h}{\alpha(2-\beta)}\left[\left\{(1-\beta)\alpha\left(T-\left(\dfrac{T}{k}+\dfrac{Q_0^{1-\beta}}{\alpha(1-\beta)k}\right)\right)\right\}^{\frac{2-\phi}{1-\beta}}\right] - A\end{array}\right] \tag{13.22}$$

Our aim is to maximize the function $Z_3(T)$, using the essential conditions for $Z_3(T)$ to be maximize is $\frac{\partial Z_3(T)}{\partial T} = 0$ and $\frac{\partial^2 Z_3}{\partial T^2} < 0$ for all T > 0.

13.4 NUMERICAL EXAMPLES

In this part, to represent theoretical outcomes just as to acquire some managerial experiences, we utilize some numerical examples to run the proposed mathematical models. We then take an observation with the support of sensitivity analysis for some significant parameters in proper units.

Example 1: For Case I

Let us consider D = 1500, h = 2.5, S = 40, P = 15, A = 250, k = 1.5. Based on the above input data and using the software MATHEMATICA 11.2.

We have the following unique optimal value for T and Z_1

$T^* = 0.632456$ and $Z_1^* = 36709.4$. See Figure 13.4 for the optimum value where horizontal axis represents time(cycle length) and vertical axis represents profit.

Example 2: For Case II

Let us consider D = 1500, h = 2.5, S = 40, P = 10, $\alpha = 25$, $\beta = 0.4$, Q = 100, A = 250, k = 1.5. Based on above input data and Using the software MATHEMATICA 11.2.

We obtain the unique optimal value for T and Z_2

$T^* = 32.033$ and $Z_2^* = 9776.48$. See Figure 13.5 for the optimum value where horizontal axis represents time (cycle length) and vertical axis represents profit.

Example 3: For Case III

Let us consider D = 1500, h = 3, S = 30, P = 10, A = 250, k = 1.5, $\alpha = 25$, $\beta = 0.4$, $Q_0 = 90$. Based on above input data and using the software MATHEMATICA 11.2.

$T^* = 17.822$ and $Z_3^* = 4246.39$. See Figure 13.6 for the optimum value where horizontal axis represents time (cycle length) and vertical axis represents profit.

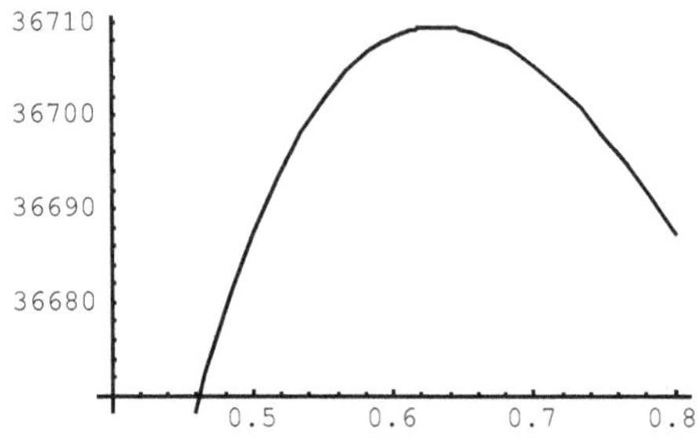

FIGURE 13.4 Total profit verses cycle length in case-I.

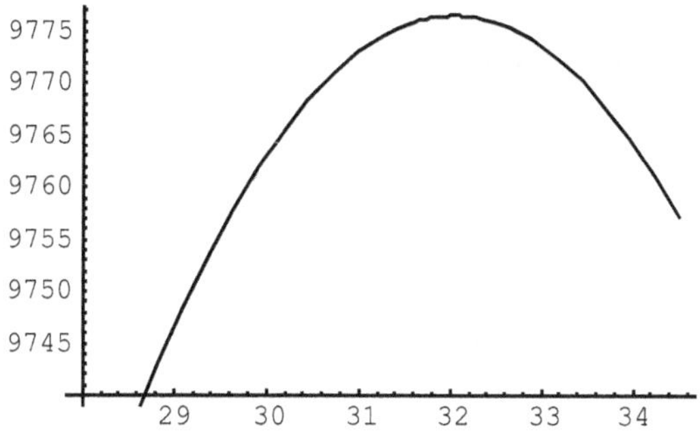

FIGURE 13.5 Total profit verses cycle length in case-II.

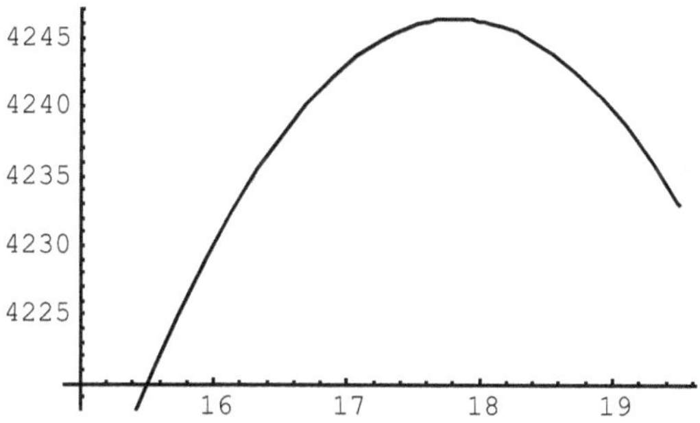

FIGURE 13.6 Total profit verses cycle length in case-III.

13.5 SENSITIVITY ANALYSIS

In today's era of the competitive business scenario, to increase the demand of any product is very much the affective role of displaying the stock of the product. Based on the data used in the above examples, the following sensitivity analysis is provided with various constraints.

Table For Case I

Parameter	Changes (in %)	T*	Z_1^*
h	−20	0.7071	36792.9
	−10	0.6666	36750
	+10	0.6030	36670.8
	+20	0.57735	36634
D	−20	0.7071	29292.9
	−10	0.6666	33000
	+10	0.6030	40420.8
	+20	0.57735	44134

Table for Case II

Parameter	Changes (in %)	T*	Z_2^*
h	−20	41.82	11139
	−10	36.40	10391.2
	+10	28.42	9262.26
	+20	25.39	8826.20
α	−20	29.91	6917.46
	−10	31.13	7140.27
	+10	32.74	11357.1
	+20	33.31	13033.7

Table for Case III

Parameter	Changes (in %)	T*	Z_3^*
h	−20	23.16	4904.88
	−10	20.19	4546.06
	+10	15.87	3991.22
	+20	14.24	3770.47
Q_0	−20	18.28	4232.43
	−10	18.04	4239.93
	+10	17.60	4251.97
	+20	17.38	4256.79

The above table reveals the following:

(i) The total profit and cycle length are both decreases with increase of holding cost ‘h’.
(ii) When the value of demand of D increases, then the profit function is also increases in case I.
(iii) When the value of α increases, then the profit function also increases in case II.
(iv) When the value of Q_0 increases, then the profit function is also increases in case III.

13.6 CONCLUSION

The proposed model delivers the optimal profit vigorously over sensitivity analysis of the parameter involved, stock-dependent demand rate which is directly connected to the present business environment. This work has assumed three types of demand based on the financial situation of any business organization: firstly, the demand rate is constant and secondly, the demand rate is in power form. Therefore, the stock-dependent demand plays a very big role in inventory management. In the chapter, we have developed an EPQ mathematical model assuming demand as a function of stock-dependent demand with three cases arises due to the stock-dependent demand rate. With the help of the software package MATHEMATICA 11.2, the numerical experimentations and sensitivity are carried out. In addition, the concavity for the proposed model profit function is demonstrated through the graph in each case. Then, the model is solved numerically to get an optimal replenishment cycle, which would maximize the total profit function. This chapter can be further extended with constraints such as permissible delay, inflation, supply chain, shortages, etc.

REFERENCES

Alfares, H. K. (2007). Inventory model with stock-level dependent demand rate and variable holding cost. *International Journal of Production Economics*, 108(1–2), 259–265.

Chung, K. J. (2003). An algorithm for an inventory model with inventory-level-dependent demand rate. *Computers & Operations Research*, 30(9), 1311–1317.

Datta, T. K., & Pal, A. K. (1990). A note on an inventory model with inventory-level-dependent demand rate. *Journal of the Operational Research Society*, 41(10), 971–975.

Gupta, K. K., Sharma, A., Singh, P. R., & Malik, A. K. (2013). Optimal ordering policy for stock-dependent demand inventory model with non-instantaneous deteriorating items. *International Journal of Soft Computing and Engineering*, 3(1), 279–281.

Kumar, S., Malik, A. K., Sharma, A., Yadav, S. K., & Singh, Y. (2016, March). *An inventory model with linear holding cost and stock-dependent demand for non-instantaneous deteriorating items.* In *AIP Conference Proceedings* (Vol. 1715, No. 1, p. 020058). AIP Publishing LLC.

Levin, R. I., Lamone, R. P., Kottas, J. F., & McLaughlin, C. P. (1972). *Production Operations Management: Contemporary Policy for Managing Operating Systems.* New York: McGraw-Hill.

Malik, A. K., Mathur, P., & Kumar, S. (2019, August). *Analysis of an inventory model with both the time dependent holding and sales revenue cost.* In *IOP Conference Series: Materials Science and Engineering* (Vol. 594, No. 1, p. 012043). IOP Publishing.

Malik, A. K., Shekhar, C., Vashisth, V., Chaudhary, A. K., & Singh, S. R. (2016, March). *Sensitivity analysis of an inventory model with non-instantaneous and time-varying deteriorating Items.* In *AIP Conference Proceedings* (Vol. 1715, No. 1, p. 020059). AIP Publishing LLC.

Mathur, P., Malik, A. K., & Kumar, S. (2019, August). *An inventory model with variable demand for non-instantaneous deteriorating products under the permissible delay in payments.* In *IOP Conference Series: Materials Science and Engineering* (Vol. 594, No. 1, p. 012042). IOP Publishing.

Rahaman, M., Mondal, S. P., Shaikh, A. A., Ahmadian, A., Senu, N., & Salahshour, S. (2020). Arbitrary-order economic production quantity model with and without deterioration: Generalized point of view. *Advances in Difference Equations*, 2020(1), 1–30.

Rout, C., Chakraborty, D., & Goswami, A. (2020). An EPQ model for deteriorating items with imperfect production, two types of inspection errors and rework under complete backordering. *International Game Theory Review*, 22(02), 2040011.

Ruidas, S., Seikh, M. R., & Nayak, P. K. (2020). An EPQ model with stock and selling price dependent demand and variable production rate in interval environment. *International Journal of System Assurance Engineering and Management*, 11, 385–399.

Sarkar, B., & Sarkar, S. (2013). Variable deterioration and demand—An inventory model. *Economic Modelling*, 31, 548–556.

Saxena P., Sharma A. & Sharma K. (2020). Partially backlogged two-warehouse ordering policy with selling price dependent demand under infinite planning horizon. *Advances and Applications in Mathematical Sciences*, 19(10), 1061–1074.

Sharma, A., Sharma, U., & Singh, C. (2017). A robust replenishment model for deteriorating items considering ramp-type demand and inflation under fuzzy environment. *International Journal of Logistics Systems and Management*, 28(3), 287–307.

Sharma, A., Sharma, U., & Singh, C. (2018). An analysis of replenishment model of deteriorating items with ramp-type demand and trade credit under the learning effect. *International Journal of Procurement Management*, 11(3), 313–342.

Shukla, H. S., Tripathi, R. P., & Sang, N. (2017). EOQ model with stock-level dependent demand and different holding cost functions. *International Journal of Operations Research and Information Systems (IJORIS)*, 8(4), 59–75.

Silver, E. A., & Peterson, R. (1985). *Decision Systems for Inventory Management and Production Planning* (Vol. 18). New York: John Wiley & Sons.

Singh, S. R., & Malik, A. K. (2011). An inventory model with stock-dependent demand with two storages capacity for non-instantaneous deteriorating items. *International Journal of Mathematical Sciences and Applications*, 1(3), 1255–1259.

Singh, S. R., Malik, A. K., & Gupta, S. K. (2011). Two warehouses inventory model for non-instantaneous deteriorating items with stock-dependent demand. *International Transactions in Applied Sciences*, 3(4), 911–920.

Teng, J. T., & Chang, C. T. (2005). Economic production quantity models for deteriorating items with price-and stock-dependent demand. *Computers & Operations Research*, 32(2), 297–308.

Vashisth V., Ajay, T., Chandra, S., & Malik, A. K. (2016). A trade credit inventory model with multivariate demand for non-instantaneous decaying products. *Indian Journal of Science and Technology*, 9(15), 1–6.

14 Models of Supply Chain Sustainability in Industrial Engineering

Abhijit Pandit and Ramakant Bhardwaj

CONTENTS

14.1 INTRODUCTION

Over the past decade, the concept of sustainability has gained traction as businesses aggressively pursue competitive advantage in a very volatile global context (Kleindorfer et al., 2005; Sharma and Henriques, 2005; Epstein, 2008). Throughout society, the term "sustainability" or "sustainable development" has become widely accepted, particularly in government (Tsai and Chou, 2008; Saha, 2009), higher education institutions (Allen-Gil et al., 2005; Fullan, 2005; Waheed et al., 2011), and corporations (Starik and Rands, 1995; Pullman et al. According to one definition, sustainability is "progress that fulfils the demands of the present without jeopardising the ability of future generations to satisfy their own requirements" (World Commission on Environment and Development, 1987, p. 8).

According to the existing literature on sustainability and supply chain management, the following research gaps have been identified: First and foremost, a comprehensive study of the literature indicated that there is a deficiency in the literature in terms of conducting a simultaneous analysis of three dimensions of sustainability (economic, environmental, and social) under the unifying banner of sustainability. Scholars noted that there is a dearth of studies in the supply chain literature that integrate all three dimensions of sustainability, and they advocated for more study to close this gap in the literature (Linton et al., 2007). There have been a number of

DOI: 10.1201/9781003089636-14

studies that have examined the three criteria of sustainability (Koplin et al., 2007; Bai and Sarkis, 2010), but the majority of these studies have been anecdotal, conceptual, or case-based in nature. Studies that have used large-scale surveys to assess the influence of sustainability efforts on performance outcomes have remained fragmented, making triangulation and generalization of data extremely difficult to achieve. As a matter of fact, there is a paucity of research that empirically analyses and confirms the implications of sustainability while also analyzing the three aspects of sustainability concurrently.

Last-mile logistics is a new study topic that has piqued the interest of both academics and practitioners in recent years, particularly during the past five years. Urbanization and population expansion are both increasing, as is the rise of e-commerce, shifts in consumer behavior, and increased emphasis being paid to environmental sustainability. Although there are many different definitions of last-mile logistics, the most widely accepted one is that it refers to the final stretch of the supply chain between the last distribution hub and the recipient's desired destination location. The final mile of the supply chain is frequently referred to as one of the most costly, inefficient, and environmentally damaging segments of the supply chain. According to some estimates, the final mile accounts for 13–75% of overall supply chain costs, depending on a variety of factors, including distance travelled. Numerous factors influence efficiency, including customer density and time windows, congestion, fragmentation of deliveries, cargo size and homogeneity, and the number of deliveries made each day. Last-mile logistics generates a variety of externalities, including greenhouse gas emissions, air pollution, noise, and traffic congestion, among others. For this reason, a more complete understanding of the final kilometer is necessary in order to improve its economic, environmental, and social viability.

Electronic commerce has had tremendous growth over the past decade, and it is expected to continue to increase in the coming years. At the moment, a rising number of conventional brick-and-mortar businesses are beginning to run an online channel. It is unavoidable that there will be rivalry between them and internet retailers that only offer items or services on the internet during this process. When it comes to same-day delivery (SDD), e-channels are accelerating the rivalry with traditional shops by providing significant convenience and virtually instant accessibility for online customers.

One practical method for conventional businesses to respond to the recent e-commerce trend is to provide same-day delivery services as well as standard shipping. Our approach is to make the decision to source fulfillment from more direct-to-consumer locations, such as their retailing storefronts, rather than from distant fulfillment facilities. There are several advantages to employing retail locations to fulfill online purchases, the most notable of which being the small distance between the retail stores and the customers. Providing a variety of services, such as store pickups, accessible return service, and so on, will benefit not only the fulfillment speed and prices but also the ability to give a variety of services. While new services are beneficial, they are not without their costs. Same-day delivery is no exception, since it increases the complexity of operations in retail businesses.

14.2 LITERATURE REVIEW

In order to better understand sustainability, nine components have been discovered and presented based on a comprehensive assessment of the literature. Following that, this study creates a research framework that illustrates the interrelationships between the components under consideration. The nine constructions are organized into four broad categories of sustainability: sustainability drivers, strategic sustainability orientation, sustainability practices, and sustainability performance. The nine constructs are organized into four categories of sustainability. The major objective is to do a thorough investigation on the long-term viability of a particular firm.

The variables included in the research framework are as follows: (a) sustainability drivers (such as external pressures and top leadership culture); (b) strategic sustainability orientation (such as economic, environmental, and social orientation); and (c) sustainability practices (such as sustainable supplier management practices, sustainable operations management practices, and sustainable customer management practices). The research framework also includes variables such as economic, environmental, and social performance. External pressures relate to the extent to which a company is subjected to the critical needs of long-term viability and survival. To put it another way, top leadership culture may be described as the extent to which top or senior management fosters an atmosphere that is proactive and devoted to long-term sustainability. Strategic social responsibility (SSO) is described as the amount to which a company is proactive and dedicated to economic, environmental, and social goals in decision-making.

This research, based on a review of prior work (Chen and Paulraj, 2004), concludes that businesses must adopt a supply chain perspective in order to successfully implement sustainability. As a result, managing suppliers, internal processes, and consumers is critical to ensuring long-term success. It is proposed in this study to examine three sets of Sustainable Supply Chain Management (SSCM) practices: sustainable supplier management practices, sustainable operations management (OM) practices, and sustainable customer management practices. Economic, environmental, and social performance are all included in the definition of sustainability performance.

14.3 RESEARCH METHODOLOGY

This study will examine the interrelationships between sustainability drivers, strategy, practices, and performance, drawing on institutional theory (DiMaggio and Powell, 1983), social contract theory, strategic orientation (Venkatraman, 1989), and the RBV of the firm, as well as other theories.

Using institutional theory, this research will investigate how external forces cause businesses to engage in SSO (H1). In this study, the theoretical perspective of attaining social legitimacy will be used to explain why companies are oriented toward sustainability under diverse constraints. Following the principles of institutional theory, this research looks into the ways in which companies alter the structure of their organizations and cultural norms (such as top leadership culture) in order to gain social legitimacy among external stakeholders such as competitors,

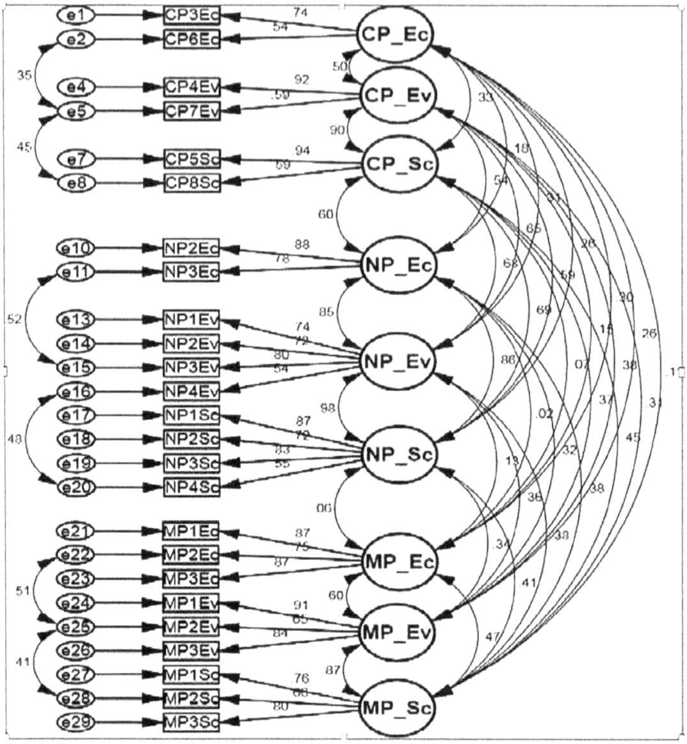

FIGURE 14.1 Research framework.

government regulators, supply chain members, and non-governmental organizations (NGOs) (H2). In the light of SCT, this study investigates how top management, via the exercise of discretion, interpretation, and perspective, builds a culture that impacts businesses' willingness to undertake sustainability efforts. (H3). By drawing on theoretical views from institutional theory and strategic choice theory, the study framework will help to validate the process of developing an organizational sustainability orientation (and, consequently, its antecedents). Furthermore, strategic sustainability orientation (SSO) is presented as a critical organization-wide perspective that enables businesses to spend a variety of resources in order to adopt economic, environmental, and social practices for sustainability (H4–H6) (H4–H6). In the end, this study investigates how sustainable supply chain strategies, such as those relating to upstream suppliers, internal operations, and downstream consumers, provide companies with a competitive advantage (H7a–H9c) by using the logic of the RBV.

14.4 HYPOTHESES TESTED

H1: Firms' perceived external pressures—that is, mimetic, coercive, and normative pressures toward sustainability—are positively related to their strategic sustainability orientation.

H2: Firms' perceived external pressures—that is, mimetic, coercive, and normative pressures toward sustainability—are positively related to their top leadership culture.

H3: Firms' proactive and committed top leadership culture is positively related to their strategic sustainability orientation.

H4: Firms' strategic sustainability orientation is positively related to their implementation of sustainable supplier management practices (i.e., supplier evaluation practices, supplier development practices, and information sharing with suppliers).

H5: Firms' strategic sustainability orientation is positively related to their implementation of sustainable operations management practices (i.e., QM and JIT practices, corporate environmental management practices, and corporate social responsibility practices).

H6: Firms' strategic sustainability orientation is positively related to their implementation of sustainable customer management practices (i.e., customer management practices and information sharing with customers).

H7: Sustainable supplier management (SSM) practices positively influence sustainability performance.

- H7a: Higher levels of adoption of SSM practices are positively related to economic performance.
- H7b: Higher levels of adoption of SSM practices are positively related to environmental performance.
- H7c: Higher levels of adoption of SSM practices are positively related to social performance.

H8: Sustainable operations management (SOM) practices positively influence sustainability performance.

- H8a: Higher levels of adoption of SOM practices are positively related to economic performance.
- H8b: Higher levels of adoption of SOM practices are positively related to environmental performance.
- H8c: Higher levels of adoption of SOM practices are positively related to social performance.

H9: Sustainable customer management practices positively influence sustainability performance.

- H9a: Higher levels of adoption of sustainable customer management practices are positively related to economic performance.

H9b: Higher levels of adoption of sustainable customer management practices are positively related to environmental performance.
H9c: Higher levels of adoption of sustainable customer management practices are positively related to social performance.

14.5 RESEARCH QUESTIONS AND OBJECTIVES

This study proposes an integrated research methodology to explore how a focus business tackles all three elements of sustainability in the supply chain in order to close research gaps. In order to do this, the following research questions have been identified:

(1) Do external constraints and the culture fostered by senior leadership have a beneficial impact on strategic sustainability orientation (SSO)?
(2) What is the impact of SSO on the supply chain management techniques employed by businesses?
(3) What supply chain management methods do companies employ in order to achieve good sustainability results?
(4) Do SSCM techniques have a beneficial impact on the results of sustainability performance?

For the purpose of answering these research questions, this chapter develops a framework that: (a) examines external pressures and top leadership culture as important predicators and drivers of a focal firm's sustainability; (b) SSO that arises in response to both external pressures and the culture created by top leadership; and (c) major constituents of SSCM practices, such as supplier evaluation/development, just-in-time delivery, and environmental sustainability.

14.5.1 Analysis of Results

On the basis of the 212 survey answers, a thorough reliability and validity evaluation of the survey instrument employed in the large-scale study was conducted. This section explains the techniques that were utilized throughout the instrument validation process, as well as the statistical data that were obtained as a consequence.

In this study, structural equation modeling (SEM) is used to evaluate the data and test hypotheses about the connections between the variables of interest. It is the perceptions of the participants that serve as the basis for the data in this study, and the hypotheses reflect a sequence of simultaneous relationships, including both external and endogenous factors. Because of the greater flexibility that depicts the interaction between data and theory, SEM methods have benefits over discriminant analysis and multiple regressions. Because the objective of this study is to explore a series of interrelationships between simultaneous endogenous and exogenous factors in constructing multidimensional constructs and examining path-dependent variances, SEM is considered an acceptable approach. Table 14.1 summarizes the statistical cut-off values for targeted CFAs that were utilized in this investigation.

TABLE 14.1
Fit Statistics for Validating the Measurement Models

Fit Statistic	Recommended Cut-Off Values
RMSEA	< 0.09
GFI	> 0.85
CFI	> 0.90
NFI	> 0.90
IFI	> 0.90
SRMR	< 0.08
AVE	> 0.50
Cronbach's alpha	> 0.60

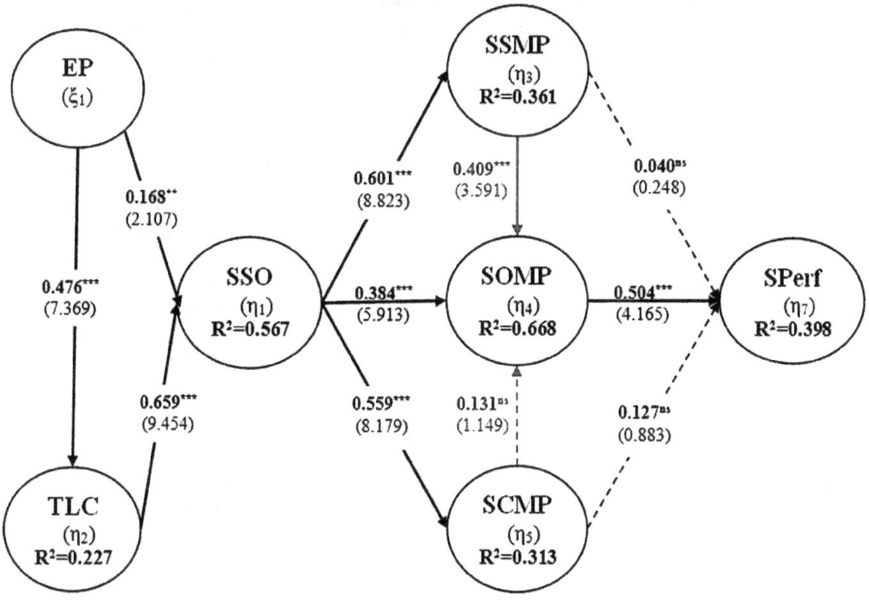

FIGURE 14.2 First-order CFA model for external pressures.

14.6 CONCLUSION

Given the paucity of empirical research on SSCM, this study offers one of the first large-scale empirical examinations into the issue. (1) Do external pressures (EPs) and the culture produced by top leadership have a beneficial impact on strategic sustainability orientation (SSO)? These are the primary research issues that this study sought to answer. (2) What is the impact of SSO on the supply chain management techniques employed by businesses? What supply chain management methods do companies employ in order to achieve good sustainability results? (4) Do SSCM techniques have a beneficial impact on the results of sustainability performance?

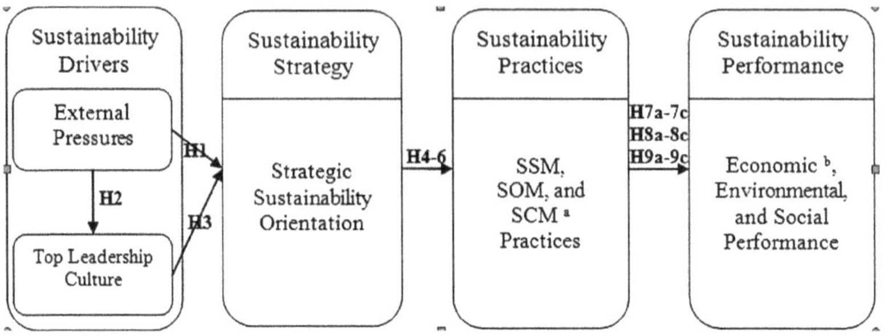

FIGURE 14.3 Revised PLS structural results. *(Sustainability Performance in Aggregation including non-significant paths).* **Note**: Bold lines are significant paths; dotted lines represent non-significant paths; t-values are in parentheses. *** significant at $p < 0.01$, ** significant at $p < 0.05$, * significant at $p < 0.10$, ns not significant. R^2 values represent the explained variance for the endogenous variables. **Paths in red represent revised paths. Legends**: EP–External Pressures, TLC–Top Leadership Culture, SSO–Strategic Sustainability Orientation, SSMP–Sustainable Supplier Management Practices, SOMP- Sustainable Operations Management Practices, SCMP–Sustainable Customer Management Practices, SPerf–Sustainability Performance (in Aggregation).

TABLE 14.2
Summary of Structural Model† Results Controlled for Social Desirability bias

Hypotheses: Path	Path Coefficient (Control Model)	t-stats. (Control Model)	R^2 a (Control Model)	Path Coefficient (Original Model)	t-stats. (Original Model)
H1: EP → SSO	0.169**	2.269	0.585	0.168**	2.208
H2: EP → TLC	0.402***	5.338	0.334	0.476***	7.111
H3: TLC → SSO	0.597***	7.446	0.585	0.659***	9.771
H4: SSO → SSMP	0.545***	5.943	0.375	0.601***	9.470
H5: SSO → SOMP	0.653***	9.474	0.507	0.704***	15.124
H6: SSO → SCMP	0.509***	6.090	0.323	0.558***	8.519
H7: SSMP → SPerf	0.006 ns	0.041	0.464	0.044 ns	0.263
H7a: SSMP → EcP	−0.045 ns	0.238	0.315	−0.012 ns	0.054
H7b: SSMP → EvP	0.113 ns	0.623	0.318	0.114 ns	0.569
H7c: SSMP → ScP	−0.056 ns	0.349	0.369	−0.026 ns	0.158
H8: SOMP → SPerf	0.420***	3.759	0.464	0.505***	3.873
H8a: SOMP → EcP	0.288**	2.187	0.315	0.419***	3.030
H8b: SOMP → EvP	0.486***	4.164	0.318	0.490***	3.709
H8c: SOMP → ScP	0.283**	2.239	0.369	0.355***	2.610
H9: SCMP → SPerf	0.112 ns	0.778	0.464	0.122 ns	0.827
H9a: SCMP → EcP	0.020 ns	0.104	0.315	0.046 ns	0.222
H9b: SCMP → EvP	−0.022 ns	0.158	0.318	−0.022 ns	0.151
H9c: SCMP → ScP	0.269**	2.025	0.369	0.280**	2.061
SDB b1: SDb →EP	−0.210ns	1.230			
SDB2: SD → TLC	−0.339ns	1.658			
SDB3: SD → SSO	−0.144ns	1.190			

(Continued)

TABLE 14.2 (Continued)

Hypotheses: Path	Path Coefficient (Control Model)	t-stats. (Control Model)	R^2 a (Control Model)	Path Coefficient (Original Model)	t-stats. (Original Model)
SDB4: SD → SSMP	−0.130ns	1.058			
SDB5: SD → SOMP	−0.118ns	1.069			
SDB6: SD → SCMP	−0.115ns	1.033			
SDB7: SD → SPerf	−0.293ns	1.573			
SDB8: SD → EcP	−0.398ns	1.769			
SDB9: SD → EvP	−0.012ns	0.143			
SDB10: SD → ScP	−0.255ns	1.599			

*significant at $p < 0.01$ or 1%, **significant at $p < 0.05$ or 5%, *significant at $p < 0.10$ or 10%, ns not significant.

R^2 values represent the explained variance for the endogenous variables.

SDB refers to social desirability bias test; SD refers to social desirability construct.

† This test is based on the *initial* structural model, not the *revised* model.

The hypotheses under consideration for these study topics are tested in 15 different ways (H1 to H9c). PLS structural equation modeling is used to test the initial predicted connections. The findings revealed that 10 hypotheses (H1 to H6, H8a to H8c, and H9c) are significant, but 5 hypotheses (H7a to H7c, as well as H9a and H9b) are found to be unsupported. A updated version of the original model was developed by estimating two additional connections in order to investigate alternate hypotheses (SSMPs to SOMPs and SCMPs to SOMPs).

Although this study supports earlier findings in those areas, it also enhances and broadens the knowledge base by focusing on the sustainable supply chain of a particular business. The first phase of the investigation involves the development of constructs for supplier management practices that measure the economic, environmental, and social dimensions of sustainability. The second phase investigates the indirect effect of these practices on operational, business, environmental, and social performance outcomes through the lens of a firm's operations management practices, as well as the direct effect of these practices on operational, business, environmental, and social performance outcomes. Second, this study indicates that lean manufacturing methods (e.g., quality management [QM] and just-in-time [JIT] practices) are antecedents to providing good advantages in operational, business, and environmental performance, as well as the ability to enhance social performance. Third, this research confirms that environmental management methods employed by businesses may contribute to the improvement of economic and environmental performance, as well as the enhancement of social performance. Fourth, this study confirms and investigates the existence of beneficial connections between corporate social responsibility practices and economic, environmental, and social performance.

14.7 APPENDIX

Glossary of Key Terms and Variables

Coercive Pressure (CP)	Sustainability-related political influences exerted by governmental regulations and/or the other firms on which your firm depends, such as important customers and parent company.
Community-oriented Outcomes (COO)	The extent to which a firm enhances the community in which it operates.
Corporate Social Involvement (CSIV)	The extent to which a firm makes philanthropic commitment within a community and to a greater society.
Corporate Sustainability Reporting (CSRP)	The extent to which a firm discloses quantitative and qualitative information on economic, environmental, and social performance.
Customer Management (CM)	The extent to which an organization manages its main customers to improve their overall satisfaction in regards to sustainability.
Economic Orientation (EcO)	The extent to which a firm is proactive and committed to positive market and financial priorities in its decision making.
Employee Motivation (EM)	The extent to which top/senior management inspires shop-floor employees to actively participate in sustainability initiatives.
Employee Wellbeing and Equity (EWE)	The extent to which a firm promotes and improves the overall quality of employees' health/safety and human rights.
Employee-oriented Outcomes (EOO)	The extent to which a firm improves the employees' well-being/equity and addresses human rights concerns.
Environmental Design Practices (EDP)	The extent to which an organization systematically integrates environmental issues into product and process design.
Environmental Management (EvM)	The extent to which a firm reduces, reuses, and recycles waste/products/energy.
Environmental Management System (EMS)	The extent to which an organization conforms to the ISO 14001 standard aimed at improving environmental performance.
Environmental Orientation (EvO)	The extent to which a firm is proactive and committed to positive ecological or green priorities in its decision-making.
Environmental Recycling Practices (ERcP)	The extent to which an organization reuses, recycles, and remanufactures materials, components, and/or returned products.
Financial Performance (FP)	The extent to which a firm achieves profit-oriented outcomes such as ROI and ROA.
Information Sharing with Customers (ISC)	The extent to which an organization receives critical and proprietary information from major customers in regards to sustainability.
Information Sharing with Suppliers (ISS)	The extent to which a firm receives critical and proprietary information from major suppliers in regards to sustainability.
Just-in-Time (JIT)	The extent to which a firm manages or streamlines the flow of production.
Managerial Attitude and Perspective (MAP)	The extent to which top or senior management views sustainability issues as opportunities rather than as threats.
Market Performance (MP)	The extent to which a firm achieves market-valued outcomes such as sales and market growth.
Mimetic Pressure (MP)	The demands that arise when your main competitors successfully adopt sustainability initiatives.
Normative Pressure (NP)	The demands that stem from collective societal expectations, such as important suppliers, labor unions, trade associations, local communities, and non-governmental organizations, with regard to sustainability.

(Continued)

Glossary of Key Terms and Variables

Operational Performance (OP)	The extent to which a firm improves outcomes in regards to cost, quality, delivery, and flexibility compared to last year's performance.
Pollution Control (PC)	The extent to which a firm reduces environmental pollution.
Quality Management (QM)	The extent to which a firm improves the quality of products/processes and maintains equipment productivity.
Social Orientation (ScO)	The extent to which a firm is proactive and committed to positive employee and communal priorities in its decision making.
Supplier Development Practices (SDP)	The extent to which a firm endeavors to improve its suppliers' performance or capabilities in regards to sustainability.
Supplier Evaluation Practices (SEP)	The extent to which a firm assesses or monitors suppliers' sustainability performance.
Supply Chain Management (SCM)	SCM is defined as the set of activities undertaken by an organization to promote effective management of its supply chain.
Sustainability	Sustainability is grounded in the concept of the triple-bottom line (TBL), a firm's simultaneous pursuit of achieving profits, preserving the planet, and enhancing society including employees.
Sustainable Supply Chain Management (SSCM)	SSCM is defined as a focal company's intra- and inter-organizational practices to manage upstream efforts, internal operations, and downstream activities in order to simultaneously achieve economic, environmental, and social performance.
Top Management Support (TMS)	The extent to which top or senior management is involved in sustainability programs.
Triple-bottom line (TBL)	The TBL defines sustainability as concurrent achievement of three objectives—economic viability, environmental stewardship, and social well-being.

REFERENCES

Allen-Gil, S., Walker, L., Thomas, G., Shevory, T., Elan, S., 2005. Forming a community partnership to enhance education in sustainability. *International Journal of Sustainability in Higher Education* 6(4), 392–402.

Bai, C., Sarkis, J., 2010. Integrating sustainability into supplier selection with grey system and rough set methodologies. *International Journal of Production Economics* 124(1), 252–264.

Chen, I.J., Paulraj, A., 2004. Towards a theory of supply chain management: The constructs and measurement. *Journal of Operations Management* 22(2), 119–150.

DiMaggio, P.J., Powell, W.W., 1983. The iron cage revised: Institutional isomorphism and collective rationality in organizational fields. *American Sociological Review* 48(2), 147–160.

Epstein, M.J., 2008. *Making sustainability work.* Sheffield, UK: Greenleaf Publishing.

Fullan, M., 2005. *Leadership & Sustainability: System Thinkers in Action.* Thousand Oaks, Calif. Corwin Press.

Koplin, J., Seuring, S., Mesterharm, M., 2007. Incorporating sustainability into supply management in the automotive industry – the case of the Volkswagen AG. *Journal of Cleaner Production* 15, 1053–1062.

Kleindorfer, P.R., Singhal, K., Van Wassenhove, L.N., 2005. Sustainable operations management. *Production and Operations Management* 14(4), 482–492.

Linton, J. D., Klassen, R., Jayaraman, V., 2007. Sustainable supply chains: An introduction. *Journal of Operations Management* 25(6), 1075–1082.

Saha, D., 2009. Empirical research on local government sustainability efforts in the USA: Gaps in the current literature. *Local Environment* 14(1), 17–30.

Sharma, S., Henriques, I., 2005. Stakeholder influences on sustainability practices in the Canadian forest products industry. *Strategic Management Journal* 26(2), 159–180.

Starik, M., Rands, G.P., 1995. Weaving an integrated web: Multilevel and multisystem perspectives of ecologically sustainable organizations. *Academy of Management Review* 20(4), 908–935.

Tsai, W., Chou, Y., 2008. Governmental policies on hydrochlorofluorocarbons (HCFCs) mitigation and its cleaner production measures – case study in Taiwan. *Journal of Cleaner Production* 16(5), 646–654.

Venkatraman, N., 1989. Strategic Orientation of business enterprises: The construct, dimensionality, and measurement. *Management Science* 35(8), 942–962.

Waheed, B., Khan, F. I., Veitch, B., Hawboldt, K., 2011. Uncertainty-based quantitative assessment of sustainability for higher education institutions. *Journal of Cleaner Production* 19(6–7), 720–732.

World Commission on Environment and Development (WCED). 1987. Our common future. Brussels. Brundtland Report.

15 Optimal Stabilization in Chaotic Maps Using Ishikawa Feedback Technique

Ashish

CONTENTS

15.1 INTRODUCTION

In the twenty-first century, due to the complex nature of chaos, nonlinear dynamics and optimization techniques have attracted considerable attention from scientists and academicians as it may be used to design secure and efficient models. Chaos has been studied over a long time in the theory of dynamical systems [1]; at present, however, it is being studied extensively in its various forms. The study of chaotic nature in deterministic dynamical systems is a key point in the current scenario of research. One of the simplest and most famous nonlinear systems which exhibits chaotic behavior is the standard logistic map [2]. The logistic map in its standard form is given by $rx(1-x)$ with $r \in [0,4]$. For extended knowledge on nonlinear dynamics, we also refer to Holmgren [3], Alligood et al. [4], Andrecut [5], Ausloos and Dirickx [6], and Devaney [7, 8].

Therefore, it is clear that chaotic nature has played an essential role in many areas of science, such as Biology, Medicine, Business, Circuits, Laser and Celestial Mechanics etc. In 1994, Wackerbauer et al. [9] proposed various complexity measures to describe the logistic map and showed that some particular measures are needed to study the special features of the logistic map. In 2013, Martinez et al. [10] analyzed the equilibrium property of chaotic system using Lyapunov exponent and

DOI: 10.1201/9781003089636-15

also studied its bifurcation diagram by using numerical simulations. Gutierrez et al. [11] further published an article on the logistic function and corresponding generalized logistic map and tried to find the exact solution of these equations. In 2014, G. C. Wu and D. Baleanu [12, 13] studied the bifurcation diagrams of the discrete fractional logistic map. Maier and Lopez [14] proposed a switching technique which is related to Parrondian games. Again in 2012, Lopez et al. [14] used this switching technique to study the complex dynamics of the extended logistic map and showed that with alternate, extended logistic map exhibits a super stable extinction solution (see also [15]).

In 2018, Ashish, Cao and Chugh [16] studied the optimization behavior of logistic map in superior orbit. Further, a superior technique was established to study the optimal dynamics of logistic system in 2019 by Ashish et al. [17]. Furthermore, in 2019, a superior method was developed to control the chaotic nature in chaotic maps [18, 19]. In this chapter, we study the stable and chaotic behavior of generalized logistic map via Ishikawa orbit. In Section 15.2, we define the Ishikawa technique and explain an empirical approach of our study. In Section 15.3, we discuss all outcomes and draw some tables to describe the results in detail. In Section 15.4, we draw a comparison table of stability range between standard Logistic map and generalized map. Finally, the chapter is summarized in Section 15.5.

15.2 PRELIMINARIES

In this section, we mention some definitions of our proposed work which are taken into account in further sections to examine the stable and chaotic behavior of one-dimensional difference equations.

Definition 15.1 For a map h defined on X, the relation given by $x_{n+1} = h(x_n)$ is known as Picard fixed-point feedback technique [3].

Definition 15.2 Let h be a map defined on X. Then, the sequence $\{x_n\}$ defined by

$$x_{n+1} = \alpha h(x_n) + (1-\alpha)x_n$$

is said to be Mann feedback technique, where $\alpha \in [0, 1]$. Notice that when $\alpha = 1$, the Ishikawa technique converts into the Picard feedback technique [20].

Definition 15.3 Let h be a map defined on X. Let x_0 be an initial point, then the Ishikawa sequence $\{x_n\}$ can be defined with the help of given function [13]:

$$x_{n+1} = \alpha h(y_n) + (1-\alpha)x_n \text{ and } y_n = \gamma h(x_n) + (1-\gamma)x_n,$$

where α and γ are belongs to open interval (0, 1) and $0 < x < 1$.

15.3 CHAOTIC MAP IN THE ISHIKAWA FEEDBACK TECHNIQUE

Here, we deal with the representation of chaotic map using the Ishikawa technique. The empirical analysis of this generalized logistic map is studied using MATLAB® to find the optimization, that is, minimum and maximum range of control parameter r for which the chaotic map exhibits optimal chaotic and stable behavior for all $x_n \in [0, 1]$. Let us define the following chaotic map:

$$x_{n+1} = r x_n \left(1 - x_n\right)^{\beta} \tag{15.1}$$

where $x_n \in [0, 1]$, $r > 0$, $n \in N$ and $\beta > 1$, as an original chaotic map. Here, the control parameter r is non-negative real number belongs to the closed interval $[r_{\min}, r_{\max}]$. Therefore, the generalized chaotic map in the Ishikawa technique is defined as:

$$x_{n+1} = \left(1-\alpha\right)x_n + \alpha\left(r y_n \left(1-y_n\right)^{\beta}\right) \text{ and } y_n = \left(1-\gamma\right)x_n + \gamma\left(r x_n \left(1-x_n\right)^{\beta}\right) \tag{15.2}$$

Then, by solving this above equation, we get

$$\begin{aligned} x_{n+1} &= \left(1-\alpha\right)x_n + \alpha r \begin{bmatrix} \left(1-\gamma\right)x_n \\ +\gamma r x_n \left(1-x_n\right)^{\beta} \end{bmatrix} \begin{bmatrix} 1-\left(1-\gamma\right)x_n \\ -\gamma r x_n \left(1-x_n\right)^{\beta} \end{bmatrix}^{\beta} \\ &= S\left(x_n, r, \alpha, \gamma, \beta\right), \end{aligned} \tag{15.3}$$

where $0 < \alpha, \gamma < 1$, $0 < x_n < 1$, $\beta > 1$ and r belongs to the interval $[r_{\min}, r_{\max}]$. This function is named as superior optimization technique. The relation is studied to measure the impact on the chaotic and stable range by changing the variables α, γ, β and x_n. In Figures 15.1–15.4, four bifurcation diagrams have been drawn for $\beta = 2, 3, 4$, and 5. Further, it is clear from the figures that as we increase the value of β in generalized chaotic map, the regime of stability increases and the regime of chaos decreases, simultaneously.

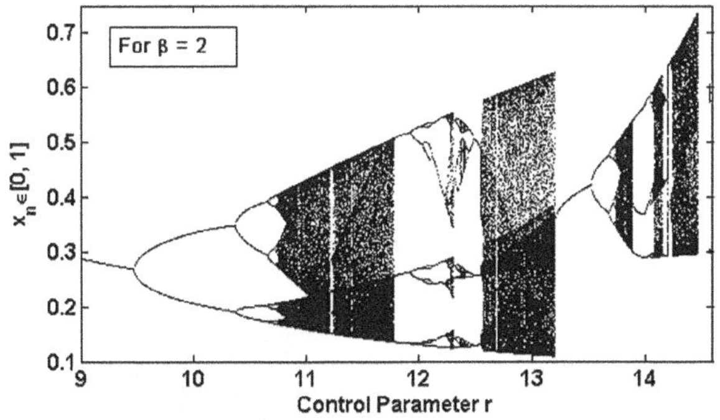

FIGURE 15.1 Bifurcation plot for the system (15.2) at $\alpha = \gamma = x_0 = 0.5$ and $\beta = 2$.

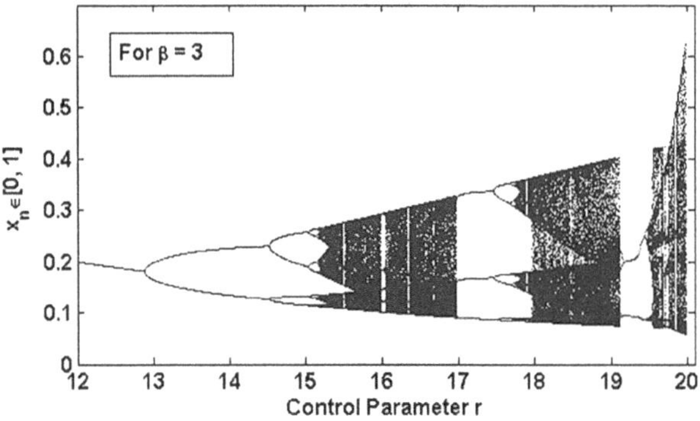

FIGURE 15.2 Bifurcation plot for the system (15.2) at $\boldsymbol{\alpha} = \boldsymbol{\gamma} = \boldsymbol{x}_0$ =0.5 and $\boldsymbol{\beta}$ = 3.

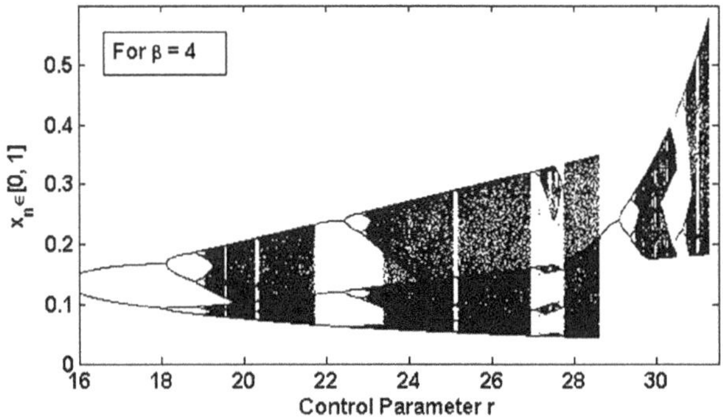

FIGURE 15.3 Bifurcation plot for the system (15.2) at $\boldsymbol{\alpha} = \boldsymbol{\gamma} = \boldsymbol{x}_0$ = 0.5 and $\boldsymbol{\beta}$ = 4.

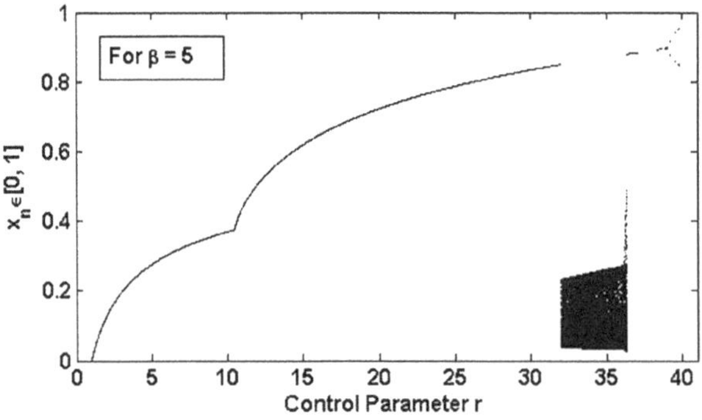

FIGURE 15.4 Bifurcation plot for the system (15.2) at $\boldsymbol{\alpha} = \boldsymbol{\gamma} = \boldsymbol{x}_0$ =0.5 and $\boldsymbol{\beta}$ = 5.

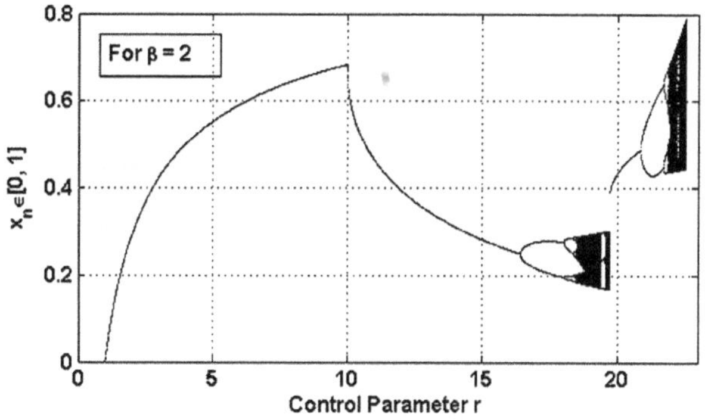

FIGURE 15.5 Bifurcation plot for the system (15.2) at $\alpha = \gamma = 0.3$, $x_0 = 0.5$ and $\beta = 2$.

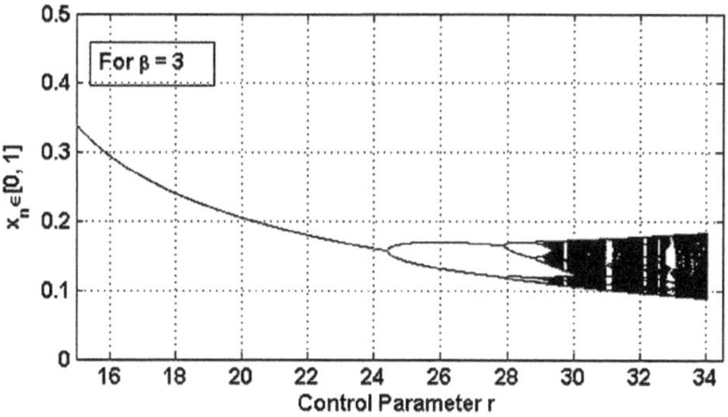

FIGURE 15.6 Bifurcation plot for the system (15.2) at $\alpha = \gamma = 0.3$, x_0 =0.5 and $\beta = 3$.

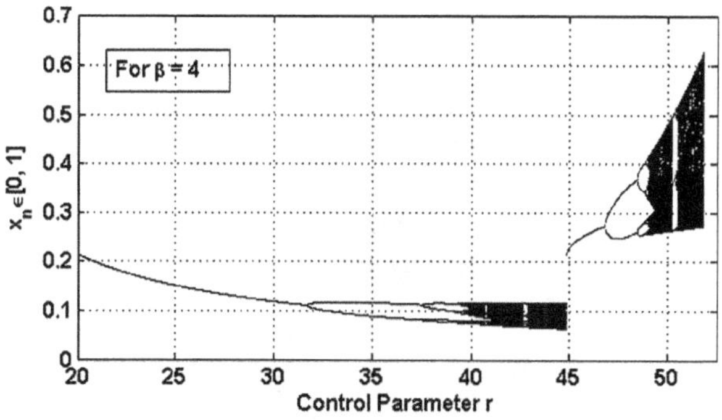

FIGURE 15.7 Bifurcation plot for the system (15.2) at $\alpha = \gamma = 0.3$, $x_0 = 0.5$ and $\beta = 4$.

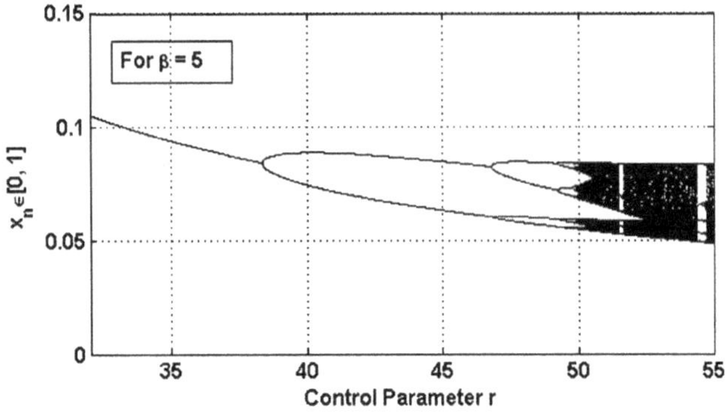

FIGURE 15.8 Bifurcation plot for the system (15.2) at $\alpha = \gamma = 0.3$, $x_0 = 0.5$ and $\beta = 5$.

Also, the bifurcation plots in Figures 15.5–15.8 represent the chaotic behavior for $\beta = 2$, 3, 4 and 5. The diagrams reveals that the chaotic regime is reduced slowly for larger values of β as the range of α and γ decreases from 1 to 0. Therefore, the generalized chaotic map through the Ishikawa technique approaches to a fully developed stable behavior for $\beta > 7$.

15.4 OPTIMAL STABILIZATION IN CHAOTIC MAPS

In above section, we have derived the definition of the Ishikawa technique and also produced its bifurcation diagrams by changing its control parameters α, γ, β and x_n. By analyzing these diagrams, we can surely say that the convergence and stability range and also the chaotic range of the control parameter r is completely associated with control parameters α, γ, β and x_n. Throughout, this section we illustrate the following results:

- For various inputs of control parameter r gives that the value of r_{max} is approximately analogues to r_{min} for all r.
- Figure 15.5–15.8 represents the bifurcation diagrams for $\beta = 2$, 3, 4, 5 and 6. The diagram shows that the chaotic regime decreases when we increased the values of the parameter β.
- Table 15.1, gives the chaotic range of the growth rate parameter r in the Ishikawa orbit. It shows that the stability range increases in the Ishikawa technique as compared to other techniques.
- The chaotic range also increases if we decrease the value of α and γ for the fixed values of β, as shown in Tables 15.2 and 15.3.
- It is observed that for x_n belongs to the closed interval [0, 0.5], the chaotic region exists for $\beta = 2$, 3, 4, 5, 6 and 7 but when x approaches to 1, chaos exists only for $\beta = 2$, as shown in Figures 15.1–15.4 and 15.5–15.8. Moreover, it is observed that the no chaos comes into existence for $\beta > 7$ and for all x_n

TABLE 15.1
Growth-rate Parameter *r* Versus Control Parameter β in Ishikawa Technique for $x_n = 0.3$ and $\alpha = \gamma = 0.9$

β	r_{min}	r_{max}
2	5.79	6.92
3	7.31	7.95
4	8.51	9.39
5	9.49	10.59
6	10.30	11.6
7	12.14	12.42

TABLE 15.2
Growth-Rate Parameter *r* Versus Control Parameter $\alpha = \gamma$ in Ishikawa Technique for $x_n = 0.3$ and $\beta = 2$

$\alpha = \gamma$	r_{min}	r_{max}
0.9	5.79	6.92
0.5	10.69	14.5
0.3	18.4	22.54

TABLE 15.3
Growth-Rate Parameter *r* Versus Control Parameter $\alpha = \gamma$ in Ishikawa Technique for $x_n = 0.3$ and $\beta = 3$

$\alpha = \gamma$	r_{min}	r_{max}
0.9	7.32	7.99
0.5	15.1	19.98
0.3	27.9	33.90

- In the remaining cases, the Ishikawa technique is either convergent to a fixed point or cyclic point but not more chaos occurs for the remaining cases.

In Table 15.1, we are taking the maximum and minimum chaotic range of parameter r. So, we try to represent the stable and cyclic nature by time series graphs for various values of β. By using equation (15.3), we evaluate the minimum range of r from where chaos starts and maximum range of the parameter r where stable behavior dies out and chaotic nature starts for various value of β and for $x_n = 0.3$, $\alpha = \gamma = 0.9$. It is analyzed that for $\beta = 2, 3, 4, 5, 6$ and 7, maximum value of r for its stable nature reaches to 5.79, 7.31, 8.51, 9.49, 10.30 and 12.14 respectively. The stability behavior of chaotic map has been shown through the time-series graphs from Figures 15.1–15.4 for different value of β which are mentioned in Table 15.1.

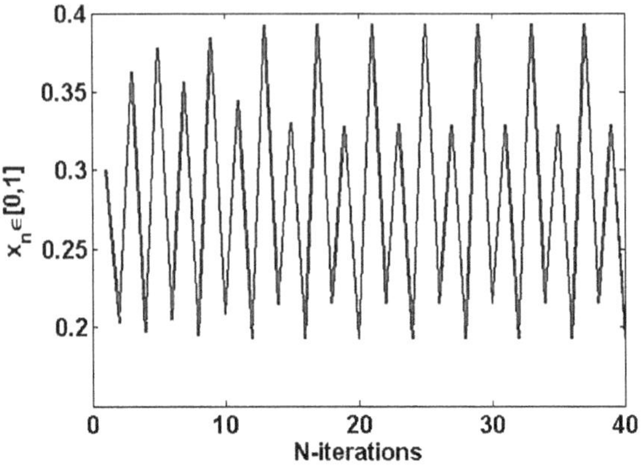

FIGURE 15.9 Cyclic behavior of the Ishikawa technique for $S(x_n, r, \alpha, \gamma, \beta)$ = (0.3, 5.7, 0.9, 0.9, 2).

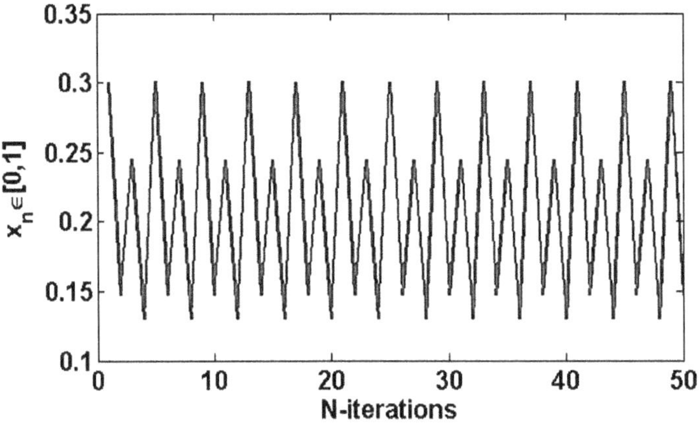

FIGURE 15.10 Cyclic behavior in the Ishikawa technique for $S(x_n, r, \alpha, \gamma, \beta)$ = (0.3, 7.3, 0.9, 0.9, 3).

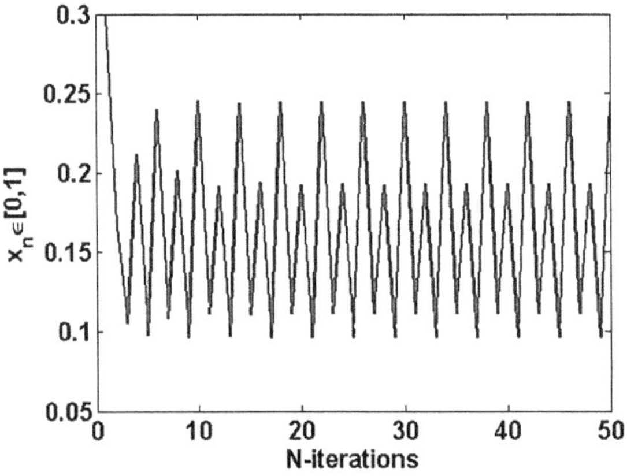

FIGURE 15.11 Cyclic behavior of generalized for $S(x_n, r, \alpha, \gamma, \beta)$= (0.3, 8.5, 0.9, 0.9, 4).

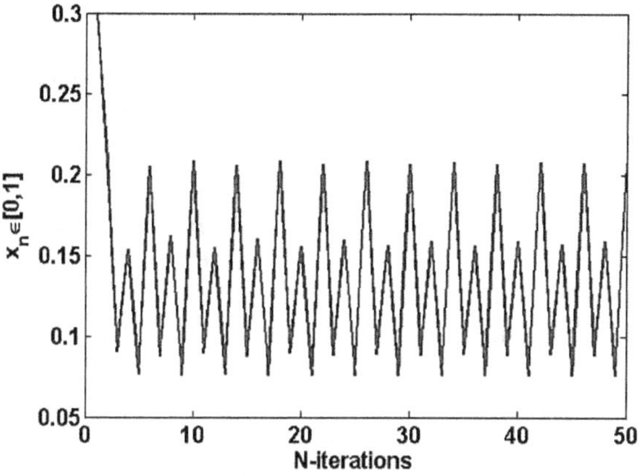

FIGURE 15.12 Cyclic behavior in the Ishikawa technique for $S(x_n, r, \alpha, \gamma, \beta)$ = (0.3, 9.5, 0.9, 0.9, 5).

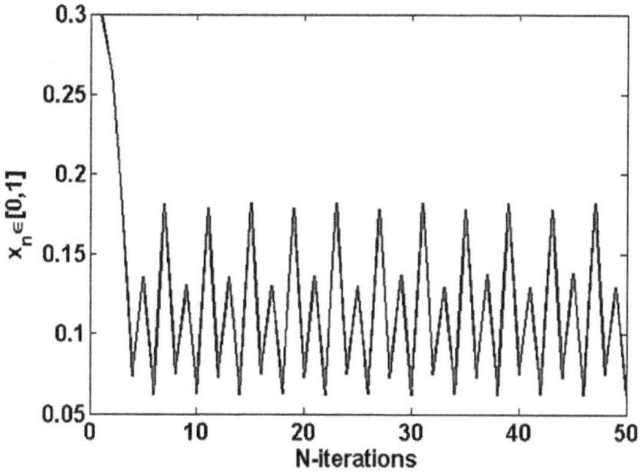

FIGURE 15.13 Cyclic behavior in the Ishikawa technique for $S(x_n, r, \alpha, \gamma, \beta)$ = (0.3, 10.3, 0.9, 0.9, 6).

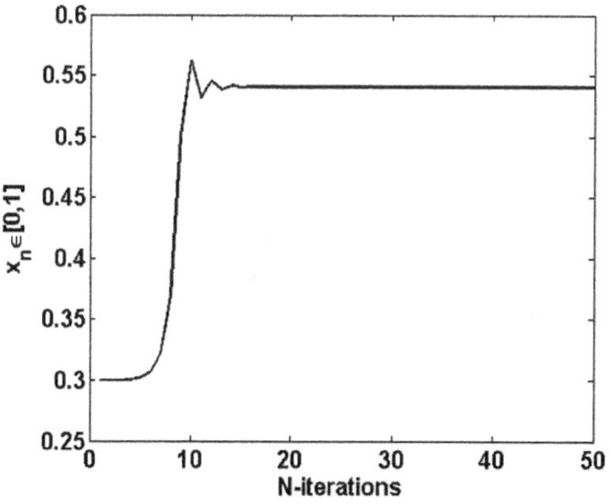

FIGURE 15.14 Cyclic behavior in the Ishikawa technique for $S(x_n, r, \alpha, \gamma, \beta)$ = (0.3, 12.1, 0.9, 0.9, 7).

15.5 CONCLUSION

In this chapter, we introduce a novel discrete time-dynamical system, which is based on the modified version of chaotic map and we call it a generalized chaotic map. The generalized map also exhibits chaotic behavior same as the standard logistic map but for larger value of r. The optimal convergence, stability and chaotic behavior of this generalized chaotic map is studied with the help of bifurcation diagrams and tables. Its stable and cyclic behavior is also studied with the help of time series graphs. Our main focus is to examine the stability and chaotic range of the generalized Logistic map via bifurcation diagram and time series graphs by changing the values of control parameters α, β and γ for all $x_n \in [0, 1]$. The range of control parameter r increases more rapidly in the generalized Logistic map when we increase the value of control parameter β for all $x_n \in [0, 1]$ and decreases the value of α and γ. Throughout the chapter, the chaotic range of generalized Logistic map depends on the freedom of three control parameters, α, β and γ, for all $x_n \in [0, 1]$.

The empirical analysis shows that the optimal range of generalized chaotic map is enhanced when we decrease the value of α and γ and increase the value of β. It can also be seen that the optimal convergence range of r in chaotic map increases very rapidly for larger values of β. Hence, it can be concluded that the chaotic regime of generalized chaotic map reduces with increasing value of β, α and γ. The period doubling bifurcation structures are plotted to examine the chaotic range of generalized chaotic map in the Ishikawa technique. By studying the bifurcation diagrams and all tables of generalized chaotic map, we observe that no chaos occurs for $\beta > 7$, for all $x_n \in [0, 1]$, $0 < \alpha$, and $\gamma < 1$. The numerical and empirical simulation shows that the range of growth rate parameter r increases as we increase the value of β; it is also observed that the range of r is much bigger than the range in standard logistic map.

REFERENCES

[1] Lorenz E. N., Deterministic nonperiodic flows. *J. Atoms. Sci.*, 20(1963), pp. 130–141.

[2] Feigenbaum M. J., Quantitative universality for a class of nonlinear transformations. *J. Stat. Phys.*, 19(1) (1978), pp. 25–52.

[3] Holmgren R. A., *A First Course in Discrete Dynamical Systems*, Springer Verlag, New York, (1994).

[4] Alligood K. T., Sauer T. D. and Yorke J. A., *Chaos: An Introduction to Dynamical Systems*, Springer Verlag, New York, (1996).

[5] Andrecut M., Logistic map as a random number generator. *Int. J. Mod. Phys. B*, 12(921) (1998), pp. 921–930.

[6] Ausloos M. and Dirickx M., *The logistic Map and the Route to Chaos, From the Beginnings to Modern Applications*. Springer, New York, (2006).

[7] Devaney R. L., *An Introduction to Chaotic Dynamical Systems*, 2nd Edition, Addison-Wesley, New York, (1948).

[8] Devaney R. L., *A First Course in Chaotic Dynamical Systems: Theory and Experiment*, Addison-Wesley, (1992).

[9] Wackerbauer R., A Comparative classification of complexity measures. *Chaos, Solitons & Fractals*, 4(1) (1994), pp. 133–173.

[10] Martinez M. G., Difference map and its electronic circuit realization. *Nonlinear Dyn.*, 74 (2013), pp. 819–830.

[11] Gutierrez M.R., A note on Verhulst's logistic equation and related logistic maps. *J. Phys. A*, 43 (2010), 5 pages.
[12] Wu G.C. and Baleanu D., *Discrete Fractional Logistic Map and Its Chaos*, Springer Science and Business Media Dordrecht, (2014), pp. 283–287.
[13] Wu G. C. and Baleanu D., Discrete Chaos in fractional delayed logistic map. *Nonlinear Dynamics*, 80 (2015), pp. 1697–1703.
[14] Maier M. P. S. and Lopez E. P., Switching induced oscillations in the logistic map. *Phys. Letters A*, 374(2010), pp. 1028–1032.
[15] Chugh R., Rani M. and Ashish, Logistic map in Noor Orbit. *Chaos Complex. Lett.* 6(3) (2012), pp. 167–175.
[16] Ashish and J. Cao, A novel fixed point feedback approaches studying the dynamical behavior of standard logistic map. *Int. J. Bifurc. Chaos*, 29(1) (2019), 16 pages.
[17] Ashish, J. Cao and R. Chugh, Chaotic behavior of logistic map in superior orbit and an improved chaos-based traffic control model. *Nonlinear Dyn.*, 94(2) (2018), pp. 959–975.
[18] Ashish and J. Cao, and R. Chugh, Controlling chaos using superior feedback technique with applications in discrete traffic models. *Int. J. Fuzzy Syst.*, 21(5) (2019), pp. 1467–1479.
[19] Ashish and J. Cao, and R. Chugh, Discrete chaotification in modulated logistic system. *Int. J. Bifurc. Chaos*, 31(5) (2021), 2150065 (14 Pages).
[20] Mann W. R., Mean value methods in iteration. *Proc. Am. Math. Soc.* 4(1953), pp. 506–510.

Index

Note: Page numbers in *italics* refers figure.

L

M

N

O

P

R

S

T

W

Y

For Product Safety Concerns and Information please contact our EU
representative GPSR@taylorandfrancis.com
Taylor & Francis Verlag GmbH, Kaufingerstraße 24, 80331 München, Germany

www.ingramcontent.com/pod-product-compliance
Lightning Source LLC
Chambersburg PA
CBHW060403220326
41598CB00023B/3005
* 9 7 8 0 3 6 7 5 4 5 4 6 8 *